T0189360

Studies in Computational Intelligence

Volume 775

The series "Studies in Computational Intelligence" (SCI) publishes new developments and advances in the various areas of computational intelligence—quickly and with a high quality. The intent is to cover the theory, applications, and design methods of computational intelligence, as embedded in the fields of engineering, computer science, physics and life sciences, as well as the methodologies behind them. The series contains monographs, lecture notes and edited volumes in computational intelligence spanning the areas of neural networks, connectionist systems, genetic algorithms, evolutionary computation, artificial intelligence, cellular automata, self-organizing systems, soft computing, fuzzy systems, and hybrid intelligent systems. Of particular value to both the contributors and the readership are the short publication timeframe and the world-wide distribution, which enable both wide and rapid dissemination of research output.

More information about this series at http://www.springer.com/series/7092

Erik Cuevas · Daniel Zaldívar
Marco Pérez-Cisneros

Advances in Metaheuristics Algorithms: Methods and Applications

Erik Cuevas
CUCEI
Universidad de Guadalajara
Guadalajara
Mexico

Marco Pérez-Cisneros
CUCEI
Universidad de Guadalajara
Guadalajara
Mexico

Daniel Zaldívar
CUCEI
Universidad de Guadalajara
Guadalajara
Mexico

ISSN 1860-949X ISSN 1860-9503 (electronic)
Studies in Computational Intelligence
ISBN 978-3-030-07736-5 ISBN 978-3-319-89309-9 (eBook)
https://doi.org/10.1007/978-3-319-89309-9

Printed on acid-free paper

This Springer imprint is published by the registered company Springer International Publishing AG part of Springer Nature
The registered company address is: Gewerbestrasse 11, 6330 Cham, Switzerland

Preface

Currently, researchers, engineers, and practitioners have faced problems of increasing complexity in several specialized areas. Some examples include mechanical design, image processing, and control processes. Such problems can be stated as optimization formulations. Under these circumstances, an objective function is defined to evaluate the quality of each candidate solution composed of the problem parameters. Then, an optimization method is used to find the candidate solution that minimizes/maximizes the objective function.

Metaheuristics is one of the most important emerging technologies of recent times for optimization proposes. Over the last years, there has been exponential growth of research activity in this field. Despite the fact that metaheuristics itself has not been precisely defined, it has become a standard term that encompasses several stochastic, population-based, and system-inspired approaches.

Metaheuristic methods use as inspiration our scientific understanding of biological, natural, or social systems, which at some level of abstraction can be represented as optimization processes. They intend to serve as general-purpose easy-to-use optimization techniques capable of reaching globally optimal or at least nearly optimal solutions. In their operation, searcher agents emulate a group of biological or social entities which interact with each other based on specialized operators that model a determined biological or social behavior. These operators are applied to a population of candidate solutions (individuals) that are evaluated with respect to an objective function. Thus, in the optimization process, individual positions are successively attracted to the optimal solution of the system to be solved.

Due to their robustness, metaheuristic techniques are well-suited options for industrial and real-world tasks. They do not need gradient information and they can operate on each kind of parameter space (continuous, discrete, combinatorial, or even mixed variants). Essentially, the credibility of metaheuristic algorithms relies on their ability to solve difficult, real-world problems with the minimal amount of human interaction.

There exist several features that clearly appear in most of the metaheuristic approaches, such as the use of diversification to force the exploration of regions of the search space, rarely visited until now, and the use of intensification or exploitation, to investigate thoroughly some promising regions. Another interesting feature is the use of memory to store the best solutions encountered.

Numerous books have been published tacking in account any of the most widely known metaheuristic methods, namely, simulated annealing, tabu search, evolutionary algorithms, ant colony algorithms, particle swarm optimization, or differential evolution, but attempts to consider the discussion of new alternative approaches are always scarce.

The excessive publication of developments based on the simple modification of popular metaheuristic methods present an important disadvantage: They avoid the opportunity to discover new techniques and procedures which can be useful to solve problems formulated by the academic and industrial communities. In the last years, several promising metaheuristic methods that consider very interesting concepts and operators have been introduced. However, they seem to have been completely overlooked in the literature, in favor of the idea of modifying, hybridizing, or restructuring popular metaheuristic approaches.

Most of the new metaheuristic algorithms present promising results. Nevertheless, they are still in their initial stage. To grow and attain their complete potential, new metaheuristic methods must be applied in a great variety of problems and contexts, so that they do not only perform well in their reported sets of optimization problems, but also in new complex formulations. The only way to accomplish this is making possible the transmission and presentation of these methods in different technical areas as optimization tools. In general, once a scientific, engineering, or practitioner recognizes a problem as a particular instance of a more generic class, he/she can select one of the different metaheuristic algorithms that guarantee an expected optimization performance. Unfortunately, the set of options are concentrated in algorithms whose popularity and high proliferation are better than the new developments.

The goal of this book is to present advances that discuss new alternative metaheuristic developments which have proved to be effective in their application to several complex problems. The book considers different new metaheuristic methods and their practical applications. This structure is important to us, because we recognize this methodology as the best way to assist researchers, lecturers, engineers, and practitioners in the solution of their own optimization problems.

This book has been structured so that each chapter can be read independently from the others. Chapter 1 describes metaheuristic methods. This chapter concentrates on elementary concepts of stochastic search. Readers that are familiar with metaheuristic algorithms may wish to skip this chapter.

In Chap. 2, a swarm algorithm, namely the Social Spider Optimization (SSO), is analyzed for solving optimization tasks. The SSO algorithm is based on the simulation of the cooperative behavior of social spiders. In SSO, individuals emulate a group of spiders which interact with each other based on the biological laws of the cooperative colony. Different to the most metaheuristic algorithms, SSO considers two different search agents (spiders): males and females. Depending on the gender, each individual is conducted by a set of different evolutionary operators which mimic the different cooperative behaviors assumed in the colony. To illustrate the proficiency and robustness of the SSO, it is compared to other well-known evolutionary methods.

Chapter 3 considers an algorithm for the optimal parameter calibration of fractional fuzzy controllers. In order to determine the best parameters, the method uses the Social Spider Optimization (SSO), which is inspired by the emulation of the collaborative behavior of social spiders. Unlike most of the existing metaheuristic algorithms, the method explicitly evades the concentration of individuals in the best positions, avoiding critical flaws such as the premature convergence to suboptimal solutions and the limited balance of exploration–exploitation. Several numerical simulations are conducted on different plants to show the effectiveness of this scheme.

In Chap. 4, the Locust Search (LS) is examined for solving some optimization tasks. The LS algorithm is based on the behavioral modeling of swarms of locusts. In LS, individuals represent a group of locusts which interact with each other based on the biological laws of the cooperative swarm. The algorithm considers two different behaviors: solitary and social. Depending on the behavior, each individual is conducted by a set of evolutionary operators which mimics different cooperative conducts that are typically found in the swarm. Different to most of existent swarm algorithms, the behavioral model in the LS approach explicitly avoids the concentration of individuals in the current best positions. Such fact allows not only to emulate in a better realistic way the cooperative behavior of the locust colony but also to incorporate a computational mechanism to avoid critical flaws that are commonly present in the popular Particle Swarm Optimization (PSO) and Differential Evolution (DE), such as the premature convergence and the incorrect exploration–exploitation balance. In order to illustrate the proficiency and robustness of the LS approach, its performance is compared to other well-known evolutionary methods. The comparison examines several standard benchmark functions which are commonly considered in the literature.

Chapter 5 presents an algorithm for parameter identification of fractional-order chaotic systems. In order to determine the parameters, the proposed method uses the metaheuristic algorithm called Locust Search (LS) which is based on the behavior of swarms of locusts. Different to the most of existent evolutionary algorithms, it explicitly avoids the concentration of individuals in the best positions, avoiding critical flaws such as the premature convergence to suboptimal solutions and the limited exploration–exploitation balance. Numerical simulations have been conducted on the fractional-order Van der Pol oscillator to show the effectiveness of the proposed scheme.

In Chap. 6, the States of Matter Search (SMS) is analyzed. The SMS algorithm is based on the modeling of the states of matter phenomenon. In SMS, individuals emulate molecules which interact with each other by using evolutionary operations based on the physical principles of the thermal energy motion mechanism. The algorithm is devised considering each state of matter one different exploration–exploitation ratio. In SMS, the evolutionary process is divided into three phases which emulate the three states of matter: gas, liquid, and solid. In each state, the evolving elements exhibit different movement capacities. Beginning from the gas state (pure exploration), the algorithm modifies the intensities of exploration and exploitation until the solid state (pure exploitation) is reached. As a result, the approach can substantially improve the balance between exploration–exploitation, yet preserving the good search capabilities of an EC method. To illustrate the proficiency and robustness of the proposed algorithm, it was compared with other well-known evolutionary methods including recent variants that incorporate diversity preservation schemas.

Although the States of Matter Search (SMS) is highly effective in locating single global optimum, it fails in providing multiple solutions within a single execution. In Chap. 7, a new multimodal optimization algorithm called the Multimodal States of Matter Search (MSMS) is introduced. Under MSMS, the original SMS is enhanced with new multimodal characteristics by means of (1) the definition of a memory mechanism to efficiently register promising local optima according to their fitness values and the distance to other probable high-quality solutions; (2) the modification of the original SMS optimization strategy to accelerate the detection of new local minima; and (3) the inclusion of a depuration procedure at the end of each state to eliminate duplicated memory elements. The performance of the proposed approach is compared to several state-of-the-art multimodal optimization algorithms considering a benchmark suite of 14 multimodal problems. The results confirm that the proposed method achieves the best balance over its counterparts regarding accuracy and computational cost.

Finally, Chap. 8 presents a methodology to implement human-knowledge-based optimization strategies. In the scheme, a Takagi–Sugeno fuzzy inference system is used to reproduce a specific search strategy generated by a human expert. Therefore, the number of rules and its configuration only depend on the expert experience without considering any learning rule process. Under these conditions, each fuzzy rule represents an expert observation that models the conditions under which candidate solutions are modified in order to reach the optimal location. To exhibit the performance and robustness of the proposed method, a comparison to other well-known optimization methods is conducted. The comparison considers several standard benchmark functions which are typically found in scientific literature. The results suggest a high performance of the proposed methodology.

This book has been structured from a teaching viewpoint. Therefore, the material is essentially directed for undergraduate and postgraduate students of Science, Engineering, or Computational Mathematics. It can be appropriate for courses such as artificial intelligence, evolutionary computation, computational intelligence, etc. Likewise, the material can be useful for researches from the evolutionary computation and artificial intelligence communities. An important propose of the book is to bridge the gap between evolutionary optimization techniques and complex engineering applications. Therefore, researchers, who are familiar with popular evolutionary computation approaches, will appreciate that the techniques discussed are beyond simple optimization tools since they have been adapted to solve significant problems that commonly arise on several engineering domains. On the other hand, students of the evolutionary computation community can prospect new research niches for their future work as master or Ph.D. thesis.

Guadalajara, Mexico
February 2018

Erik Cuevas
Daniel Zaldívar
Marco Pérez-Cisneros

Contents

Chapter 1
Introduction

This chapter provides a basic introduction to optimization methods, defining their main characteristics. This chapter provides a basic introduction to optimization methods, defining their main characteristics. The main objective of this chapter is to present to metaheuristic methods as alternative approaches for solving optimization problems. The study of the optimization methods is conducted in such a way that it is clear the necessity of using metaheuristic methods for the solution of engineering problems.

1.1 Definition of an Optimization Problem

The vast majority of image processing and pattern recognition algorithms use some form of optimization, as they intend to find some solution which is "best" according to some criterion. From a general perspective, an optimization problem is a situation that requires to decide for a choice from a set of possible alternatives in order to reach a predefined/required benefit at minimal costs [1].

Consider a public transportation system of a city, for example. Here the system has to find the "best" route to a destination location. In order to rate alternative solutions and eventually find out which solution is "best," a suitable criterion has to be applied. A reasonable criterion could be the distance of the routes. We then would expect the optimization algorithm to select the route of shortest distance as a solution. Observe, however, that other criteria are possible, which might lead to different "optimal" solutions, e.g., number of transfers, ticket price or the time it takes to travel the route leading to the fastest route as a solution.

Mathematically speaking, optimization can be described as follows: Given a function $f : S \to \mathbb{R}$ which is called the objective function, find the argument which minimizes f:

E. Cuevas et al., *Advances in Metaheuristics Algorithms: Methods and Applications*, Studies in Computational Intelligence 775,
https://doi.org/10.1007/978-3-319-89309-9_1

$$x^* = \arg\min_{x \in S} f(x) \quad (1.1)$$

S defines the so-called solution set, which is the set of all possible solutions for the optimization problem. Sometimes, the unknown(s) x are referred to design variables. The function f describes the optimization criterion, i.e., enables us to calculate a quantity which indicates the "quality" of a particular x.

In our example, S is composed by the subway trajectories and bus lines, etc., stored in the database of the system, x is the route the system has to find, and the optimization criterion $f(x)$ (which measures the quality of a possible solution) could calculate the ticket price or distance to the destination (or a combination of both), depending on our preferences.

Sometimes there also exist one or more additional constraints which the solution x^* has to satisfy. In that case we talk about constrained optimization (opposed to unconstrained optimization if no such constraint exists). As a summary, an optimization problem has the following components:

- One or more design variables x for which a solution has to be found
- An objective function $f(x)$ describing the optimization criterion
- A solution set S specifying the set of possible solutions x
- (optional) one or more constraints on x.

In order to be of practical use, an optimization algorithm has to find a solution in a reasonable amount of time with reasonable accuracy. Apart from the performance of the algorithm employed, this also depends on the problem at hand itself. If we can hope for a numerical solution, we say that the problem is well-posed. For assessing whether an optimization problem is well-posed, the following conditions must be fulfilled:

1. A solution exists.
2. There is only one solution to the problem, i.e., the solution is unique.
3. The relationship between the solution and the initial conditions is such that small perturbations of the initial conditions result in only small variations of x^*.

1.2 Classical Optimization

Once a task has been transformed into an objective function minimization problem, the next step is to choose an appropriate optimizer. Optimization algorithms can be divided in two groups: derivative-based and derivative-free [2].

In general, $f(x)$ may have a nonlinear form respect to the adjustable parameter x. Due to the complexity of $f(\cdot)$, in classical methods, it is often used an iterative algorithm to explore the input space effectively. In iterative descent methods, the next point x_{k+1} is determined by a step down from the current point x_k in a direction vector $\mathbf{d}$:

$$x_{k+1} = x_k + \alpha \mathbf{d}, \tag{1.2}$$

where α is a positive step size regulating to what extent to proceed in that direction. When the direction d in Eq. 1.1 is determined on the basis of the gradient (**g**) of the objective function $f(\cdot)$, such methods are known as gradient-based techniques.

The method of steepest descent is one of the oldest techniques for optimizing a given function. This technique represents the basis for many derivative-based methods. Under such a method, the Eq. 1.3 becomes the well-known gradient formula:

$$x_{k+1} = x_k - \alpha\, \mathbf{g}(f(x)), \tag{1.3}$$

However, classical derivative-based optimization can be effective as long the objective function fulfills two requirements:

- The objective function must be two-times differentiable.
- The objective function must be uni-modal, i.e., have a single minimum.

A simple example of a differentiable and uni-modal objective function is

$$f(x_1, x_2) = 10 - e^{-\left(x_1^2 + 3 \cdot x_2^2\right)} \tag{1.4}$$

Figure 1.1 shows the function defined in Eq. 1.4.

Unfortunately, under such circumstances, classical methodsare only applicable for a few types of optimization problems. For combinatorial optimization, there is no definition of differentiation.

Furthermore, there are many reasons why an objective function might not be differentiable. For example, the "floor" operation in Eq. 1.5 quantizes the function in Eq. 1.4, transforming Fig. 1.1 into the stepped shape seen in Fig. 1.2. At each step's edge, the objective function is non-differentiable:

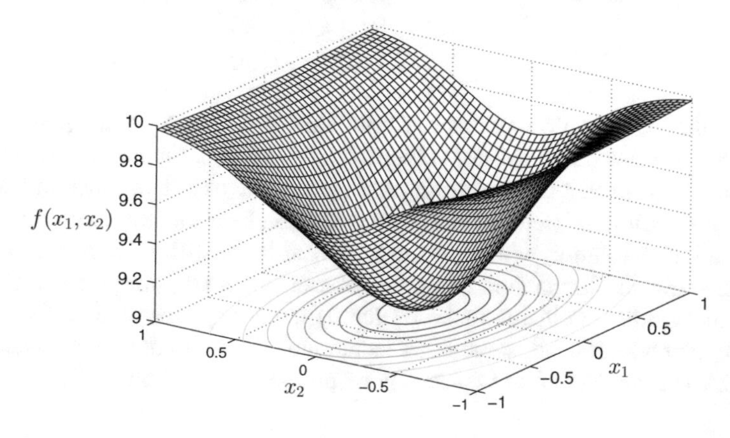

Fig. 1.1 Uni-modal objective function

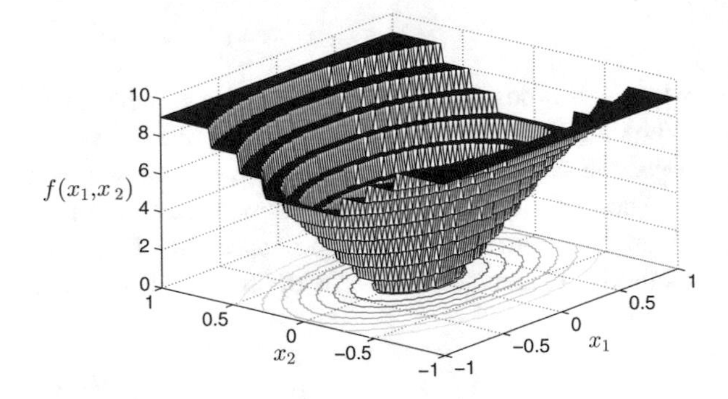

Fig. 1.2 A non-differentiable, quantized, uni-modal function

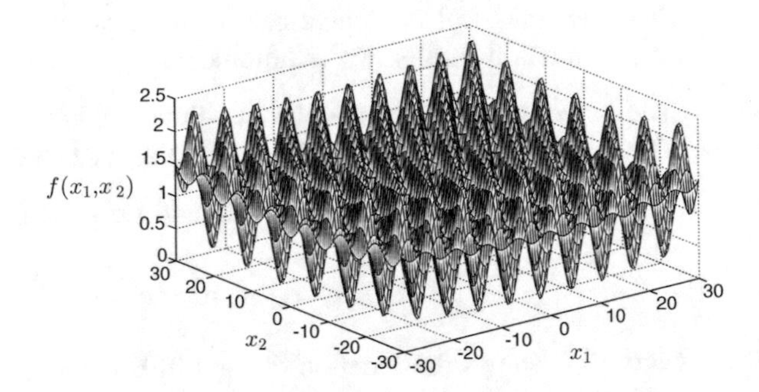

Fig. 1.3 The Griewank multi-modal function

$$f(x_1,x_2) = \text{floor}\left(10 - e^{-\left(x_1^2+3\cdot x_2^2\right)}\right) \tag{1.5}$$

Even in differentiable objective functions, gradient-based methods might not work. Let us consider the minimization of the Griewank function as an example.

$$\begin{aligned} \text{minimize} \quad & f(x_1,x_2) = \frac{x_1^2+x_2^2}{4000} - \cos(x_1)\cos\left(\frac{x_2}{\sqrt{2}}\right) + 1 \\ \text{subject to} \quad & -30 \le x_1 \le 30 \\ & -30 \le x_2 \le 30 \end{aligned} \tag{1.6}$$

From the optimization problem formulated in Eq. 1.6, it is quite easy to understand that the global optimal solution is $x_1 = x_2 = 0$. Figure 1.3 visualizes the function defined in Eq. 1.6. According to Fig. 1.3, the objective function has many local optimal solutions (multimodal) so that the gradient methods with a randomly generated initial solution will converge to one of them with a large probability.

Considering the limitations of gradient-based methods, image processing and pattern recognition problems make difficult their integration with classical optimization methods. Instead, some other techniques which do not make assumptions and which can be applied to wide range of problems are required [3].

1.3 Metaheuristic Computation Methods

Metaheuristic computation (EC) [4] methods are derivative-free procedures, which do not require that the objective function must be neither two-times differentiable nor uni-modal. Therefore, metaheuristic methods as global optimization algorithms can deal with non-convex, nonlinear, and multimodal problems subject to linear or nonlinear constraints with continuous or discrete decision variables.

The field of EC has a rich history. With the development of computational devices and demands of industrial processes, the necessity to solve some optimization problems arose despite the fact that there was not sufficient prior knowledge (hypotheses) on the optimization problem for the application of a classical method. In fact, in the majority of image processing and pattern recognition cases, the problems are highly nonlinear, or characterized by a noisy fitness, or without an explicit analytical expression as the objective function might be the result of an experimental or simulation process. In this context, the metaheuristic methods have been proposed as optimization alternatives.

An EC technique is a general method for solving optimization problems. It uses an objective function in an abstract and efficient manner, typically without utilizing deeper insights into its mathematical properties. metaheuristic methods do not require hypotheses on the optimization problem nor any kind of prior knowledge on the objective function. The treatment of objective functions as "black boxes" [5] is the most prominent and attractive feature of metaheuristic methods.

Metaheuristic methods obtain knowledge about the structure of an optimization problem by utilizing information obtained from the possible solutions (i.e., candidate solutions) evaluated in the past. This knowledge is used to construct new candidate solutions which are likely to have a better quality.

Recently, several metaheuristic methods have been proposed with interesting results. Such approaches uses as inspiration our scientific understanding of biological, natural or social systems, which at some level of abstraction can be represented as optimization processes [6]. These methods include the social behavior of bird flocking and fish schooling such as the Particle Swarm Optimization (PSO) algorithm [7], the cooperative behavior of bee colonies such as the Artificial Bee Colony (ABC) technique [8], the improvisation process that occurs when a musician searches for a better state of harmony such as the Harmony Search (HS) [9], the emulation of the bat behavior such as the Bat Algorithm (BA) method [10], the mating behavior of firefly insects such as the Firefly (FF) method [11], the social-spider behavior such as the Social Spider Optimization (SSO) [12], the simulation of the animal behavior in a group such as the Collective Animal Behavior [13], the emulation of immunological systems as the clonal selection algorithm (CSA) [14], the simulation of the electromagnetism phenomenon as the electromagnetism-Like algorithm [15], and the emulation of the differential and conventional evolution in species such as the Differential Evolution (DE) [16] and Genetic Algorithms (GA) [17], respectively.

1.3.1 Structure of a Metaheuristic Computation Algorithm

From a conventional point of view, an EC method is an algorithm that simulates at some level of abstraction a biological, natural or social system. To be more specific, a standard EC algorithm includes:

1. One or more populations of candidate solutions are considered.
2. These populations change dynamically due to the production of new solutions.
3. A fitness function reflects the ability of a solution to survive and reproduce.
4. Several operators are employed in order to explore an exploit appropriately the space of solutions.

The metaheuristic methodology suggest that, on average, candidate solutions improve their fitness over generations (i.e., their capability of solving the optimization problem). A simulation of the evolution process based on a set of candidate solutions whose fitness is properly correlated to the objective function to optimize will, on average, lead to an improvement of their fitness and thus steer the simulated population towards the global solution.

Most of the optimization methods have been designed to solve the problem of finding a global solution of a nonlinear optimization problem with box constraints in the following form:

$$\begin{array}{lll} \text{maximize} & f(\mathbf{x}), & \mathbf{x} = (x_1, \ldots, x_d) \in \mathbb{R}^d \\ \text{subject to} & & \mathbf{x} \in \mathbf{X} \end{array} \tag{1.7}$$

where $f : \mathbb{R}^d \rightarrow \mathbb{R}$ is a nonlinear function whereas $\mathbf{X} = \{\mathbf{x} \in \mathbb{R}^d | l_i \leq x_i \leq u_i, i = 1, \ldots, d.\}$ is a bounded feasible search space, constrained by the lower (l_i) and upper (u_i) limits.

In order to solve the problem formulated in Eq. 1.6, in an Metaheuristic computation method, a population $\mathbf{p}^k(\{\mathbf{p}_1^k, \mathbf{p}_2^k, \ldots, \mathbf{p}_N^k\})$ of N candidate solutions (individuals) evolves from the initial point ($k = 0$) to a total *gen* number iterations ($k = gen$). In its initial point, the algorithm begins by initializing the set of N candidate solutions with values that are randomly and uniformly distributed between the pre-specified lower (l_i) and upper (u_i) limits. In each iteration, a set of metaheuristic operators are applied over the population $\mathbf{P}^k$ to build the new population $\mathbf{P}^{k+1}$. Each candidate solution $\mathbf{p}_i^k$ $(i \in [1, \ldots, N])$ represents a d-dimensional vector $\left\{p_{i,1}^k, p_{i,2}^k, \ldots, p_{i,d}^k\right\}$ where each dimension corresponds to a decision variable of the optimization problem at hand. The quality of each candidate solution $\mathbf{p}_i^k$ is evaluated by using an objective function $f(\mathbf{p}_i^k)$ whose final result represents the fitness value of $\mathbf{p}_i^k$. During the evolution process, the best candidate solution $\mathbf{g}$ $(g_1, g_2, \ldots, g_d)$ seen so-far is preserved considering that it represents the best available solution. Figure 1.4 presents a graphical representation of a basic cycle of a metaheuristic method.

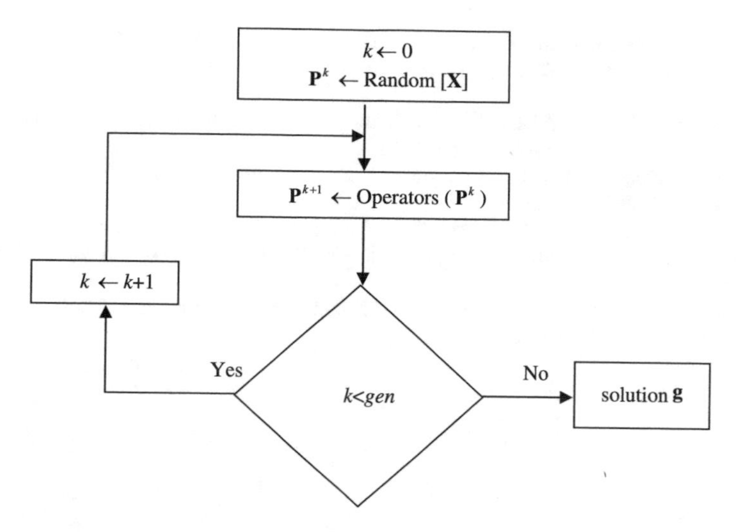

Fig. 1.4 The basic cycle of a metaheuristic method

References

1. Akay, B., Karaboga, D.: A survey on the applications of artificial bee colony in signal, image, and video processing. SIViP **9**(4), 967–990 (2015)
2. Yang, X.-S.: Engineering Optimization. Wiley, USA (2010)
3. Treiber, M.A.: Optimization for Computer Vision: An Introduction to Core Concepts and Methods. Springer, Berlin (2013)
4. Simon, D.: Evolutionary Optimization Algorithms. Wiley, USA (2013)
5. Blum, C., Roli, A.: Metaheuristics in combinatorial optimization: overview and conceptual comparison. ACM Comput. Surv. (CSUR) **35**(3), 268–308 (2003). https://doi.org/10.1145/937503.937505
6. Nanda, S.J., Panda, G.: A survey on nature inspired metaheuristic algorithms for partitional clustering. Swarm Evol. Comput. **16**, 1–18 (2014)
7. Kennedy, J., Eberhart, R.: Particle swarm optimization. In: Proceedings of the 1995 IEEE International Conference on Neural Networks, vol. 4, pp. 1942–1948, December 1995
8. Karaboga, D.: An idea based on honey bee swarm for numerical optimization. Technical Report-TR06. Engineering Faculty, Computer Engineering Department, Erciyes University (2005)
9. Geem, Z.W., Kim, J.H., Loganathan, G.V.: A new heuristic optimization algorithm: harmony search. Simulations **76**, 60–68 (2001)
10. Yang, X.S.: A new metaheuristic bat-inspired algorithm. In: Cruz, C., González, J., Krasnogor, G.T.N., Pelta, D.A. (eds.) Nature Inspired Cooperative Strategies for Optimization (NISCO 2010), Studies in Computational Intelligence, vol. 284, pp. 65–74. Springer, Berlin (2010)
11. Yang, X.S.: Firefly algorithms for multimodal optimization. In: Stochastic Algorithms: Foundations and Applications, SAGA 2009. Lecture Notes in Computer Sciences, vol. 5792, pp. 169–178 (2009)
12. Cuevas, E., Cienfuegos, M., Zaldívar, D., Pérez-Cisneros, M.: A swarm optimization algorithm inspired in the behavior of the social-spider. Expert Syst. Appl. **40**(16), 6374–6384 (2013)

13. Cuevas, E., González, M., Zaldivar, D., Pérez-Cisneros, M., García, G.: An algorithm for global optimization inspired by collective animal behaviour. Discrete Dyn. Nat. Soc. art. no. 638275 (2012)
14. de Castro, L.N., von Zuben, F.J.: Learning and optimization using the clonal selection principle. IEEE Trans. Evol. Comput. **6**(3), 239–251 (2002)
15. Birbil, Ş.I., Fang, S.C.: An electromagnetism-like mechanism for global optimization. J. Glob. Optim. **25**(1), 263–282 (2003)
16. Storn, R., Price, K.: Differential evolution—a simple and efficient adaptive scheme for global optimisation over continuous spaces. Technical Report TR-95–012. ICSI, Berkeley, CA (1995)
17. Goldberg, D.E.: Genetic Algorithm in Search Optimization and Machine Learning. Addison-Wesley, Boston (1989)

Chapter 2
The Metaheuristic Algorithm of the Social-Spider

Metaheuristic is a computer science field which emulates the cooperative behavior of natural systems such as insects or animals. Many methods resulting from these models have been suggested to solve several complex optimization problems. In this chapter, a metaheuristic approach known as the Social Spider Optimization (SSO) is analyzed for solving optimization problems. The SSO method considers the simulation of the collective operation of social-spiders. In SSO, candidate solutions represent a set of spiders which interacts among them based on the natural laws of the colony. The algorithm examines two different kinds of search agents (spiders): males and females. According to the gender, each element is conducted by a set of different operations which imitate different behaviors that are commonly observed in the colony.

2.1 Introduction

The collective intelligent behavior of insect or animal groups in nature such as flocks of birds, colonies of ants, schools of fish, swarms of bees and termites have attracted the attention of researchers. The aggregative conduct of insects or animals is known as swarm behavior. Entomologists have studied this collective phenomenon to model biological groups in nature while engineers have applied these models as a framework for solving complex real-world problems. This branch of artificial intelligence which deals with the collective behavior of elements through complex interaction of individuals with no supervision is frequently addressed as swarm intelligence. Bonabeau defined swarm intelligence as "any attempt to design algorithms or distributed problem solving devices inspired by the collective behavior of the social insect colonies and other animal societies" [1]. Swarm intelligence has some advantages such as scalability, fault tolerance, adaptation, speed, modularity, autonomy and parallelism [2].

E. Cuevas et al., *Advances in Metaheuristics Algorithms: Methods and Applications*, Studies in Computational Intelligence 775,
https://doi.org/10.1007/978-3-319-89309-9_2

The key components of swarm intelligence are self-organization and labor division. In a self-organizing system, each of the covered units responds to local stimuli individually and may act together to accomplish a global task, via a labor separation which avoids a centralized supervision. The entire system can thus efficiently adapt to internal and external changes.

Several metaheuristic algorithms have been developed by a combination of deterministic rules and randomness, mimicking the behavior of insect or animal groups in nature. Such methods include the social behavior of bird flocking and fish schooling such as the Particle Swarm Optimization (PSO) algorithm [3], the cooperative behavior of bee colonies such as the Artificial Bee Colony (ABC) technique [4], the social foraging behavior of bacteria such as the Bacterial Foraging Optimization Algorithm (BFOA) [5], the simulation of the herding behavior of krill individuals such as the Krill Herd (KH) method [6], the mating behavior of firefly insects such as the Firefly (FF) method [7] and the emulation of the lifestyle of cuckoo birds such as the Cuckoo Optimization Algorithm (COA) [8].

In particular, insect colonies and animal groups provide a rich set of metaphors for designing metaheuristic optimization algorithms. Such cooperative entities are complex systems that are composed by individuals with different cooperative-tasks where each member tends to reproduce specialized behaviors depending on its gender [9]. However, most of metaheuristic algorithms model individuals as unisex entities that perform virtually the same behavior. Under such circumstances, algorithms waste the possibility of adding new and selective operators as a result of considering individuals with different characteristics such as sex, task-responsibility, etc. These operators could incorporate computational mechanisms to improve several important algorithm characteristics including population diversity and searching capacities.

Although PSO and ABC are the most popular metaheuristic algorithms for solving complex optimization problems, they present serious flaws such as premature convergence and difficulty to overcome local minima [10, 11]. The cause for such problems is associated to the operators that modify individual positions. In such algorithms, during their evolution, the position of each agent for the next iteration is updated yielding an attraction towards the position of the best particle seen so-far (in case of PSO) or towards other randomly chosen individuals (in case of ABC). As the algorithm evolves, those behaviors cause that the entire population concentrates around the best particle or diverges without control. It does favors the premature convergence or damage the exploration–exploitation balance [12, 13].

The interesting and exotic collective behavior of social insects have fascinated and attracted researchers for many years. The collaborative swarming behavior observed in these groups provides survival advantages, where insect aggregations of relatively simple and "unintelligent" individuals can accomplish very complex tasks using only limited local information and simple rules of behavior [14]. Social-spiders are a representative example of social insects [15]. A social-spider is a spider species whose members maintain a set of complex cooperative behaviors [16]. Whereas most spiders are solitary and even aggressive toward other members of their own species, social-spiders show a tendency to live in groups, forming long-lasting aggregations often referred to as colonies [17]. In a social-spider colony, each member, depending on its gender, executes a variety of tasks such as predation, mating, web design, and social interaction [17, 18]. The web it is an important part of the colony because it is not only used as a common environment for all members, but also as a communication channel among them [19] Therefore, important information (such as trapped prays or mating possibilities) is transmitted by small vibrations through the web. Such information, considered as a local knowledge, is employed by each member to conduct its own cooperative behavior, influencing simultaneously the social regulation of the colony [20].

In this chapter, a metaheuristic algorithm, called the Social Spider Optimization (SSO) is analyzed for solving optimization tasks. The SSO algorithm is based on the simulation of the cooperative behavior of social-spiders. In this algorithm, individuals emulate a group of spiders which interact to each other based on the biological laws of the cooperative colony. The algorithm considers two different search agents (spiders): males and females. Depending on gender, each individual is conducted by a set of different evolutionary operators which mimic different cooperative behaviors that are typical in a colony. Different to most of existent metaheuristic algorithms, in the approach, each individual is modeled considering two genders. Such fact allows not only to emulate in a better realistic way the cooperative behavior of the colony, but also to incorporate computational mechanisms to avoid critical flaws commonly present in the popular PSO and ABC algorithms, such as the premature convergence and the incorrect exploration–exploitation balance. In order to illustrate the proficiency and robustness of the approach, it is compared to other well-known evolutionary methods. The comparison examines several standard benchmark functions which are commonly considered in the literature. The results show a high performance of the method for searching a global optimum in several benchmark functions.

This chapter is organized as follows. In Sect. 2.2, we introduce basic biological aspects of the algorithm. In Sect. 2.3, the novel SSO algorithm and its characteristics are both described. Section 2.4 presents the experimental results and the comparative study. Finally, in Sect. 2.5, conclusions are drawn.

2.2 Biological Concepts

Social insect societies are complex cooperative systems that self-organize within a set of constraints. Cooperative groups are better at manipulating and exploiting their environment, defending resources and brood, and allowing task specialization among group members [21, 22]. A social insect colony functions as an integrated unit that not only possesses the ability to operate at a distributed manner, but also to undertake enormous construction of global projects [23]. It is important to acknowledge that global order in social insects can arise as a result of internal interactions among members.

A few species of spiders have been documented exhibiting a degree of social behavior [15]. The behavior of spiders can be generalized into two basic forms: solitary spiders and social spiders [17]. This classification is made based on the level of cooperative behavior that they exhibit [18]. In one side, solitary spiders create and maintain their own web while live in scarce contact to other individuals of the same species. In contrast, social spiders form colonies that remain together over a communal web with close spatial relationship to other group members [19].

A social spider colony is composed of two fundamental components: its members and the communal web. Members are divided into two different categories: males and females. An interesting characteristic of social-spiders is the highly female-biased population. Some studies suggest that the number of male spiders barely reaches the 30% of the total colony members [17, 24]. In the colony, each member, depending on its gender, cooperate in different activities such as building and maintaining the communal web, prey capturing, mating and social contact [20]. Interactions among members are either direct or indirect [25]. Direct interactions imply body contact or the exchange of fluids such as mating. For indirect interactions, the communal web is used as a "medium of communication" which conveys important information that is available to each colony member [19]. This information encoded as small vibrations is a critical aspect for the collective coordination among members [20]. Vibrations are employed by the colony members to decode several messages such as the size of the trapped preys, characteristics of the neighboring members, etc. The intensity of such vibrations depend on the weight and distance of the spiders that have produced them.

In spite of the complexity, all the cooperative global patterns in the colony level are generated as a result of internal interactions among colony members [26]. Such internal interactions involve a set of simple behavioral rules followed by each spider in the colony. Behavioral rules are divided into two different classes: social interaction (cooperative behavior) and mating [27].

As a social insect, spiders perform cooperative interaction with other colony members. The way in which this behavior takes place depends on the spider gender. Female spiders which show a major tendency to socialize present an attraction or dislike over others, irrespectively of gender [17]. For a particular female spider, such attraction or dislike is commonly developed over other spiders according to their vibrations which are emitted over the communal web and represent strong colony members [20]. Since the vibrations depend on the weight and distance of the members which provoke them, stronger vibrations are produced either by big spiders or neighboring members [19]. The bigger a spider is, the better it is considered as a colony member. The final decision of attraction or dislike over a determined member is taken according to an internal state which is influenced by several factors such as reproduction cycle, curiosity and other random phenomena [20].

Different to female spiders, the behavior of male members is reproductive-oriented [28]. Male spiders recognize themselves as a subgroup of alpha males which dominate the colony resources. Therefore, the male population is divided into two classes: dominant and non-dominant male spiders [28]. Dominant male spiders have better fitness characteristics (normally size) in comparison to non-dominant. In a typical behavior, dominant males are attracted to the closest female spider in the communal web. In contrast, non-dominant male spiders tend to concentrate upon the center of the male population as a strategy to take advantage of the resources wasted by dominant males [29].

Mating is an important operation that no only assures the colony survival, but also allows the information exchange among members. Mating in a social-spider colony is performed by dominant males and female members [30]. Under such circumstances, when a dominant male spider locates one or more female members within a specific range, it mates with all the females in order to produce offspring [31].

2.3 The SSO Algorithm

In this chapter, the operational principles from the social-spider colony have been used as guidelines for developing a new metaheuristic optimization algorithm. The SSO assumes that entire search space is a communal web, where all the social-spiders interact to each other. In the approach, each solution within the search space represents a spider position in the communal web. Every spider receives a weight according to the fitness value of the solution that is symbolized by the social-spider. The algorithm models two different search agents (spiders): males and females. Depending on gender, each individual is conducted by a set of different evolutionary operators which mimic different cooperative behaviors that are commonly assumed within the colony.

An interesting characteristic of social-spiders is the highly female-biased populations. In order to emulate this fact, the algorithm starts by defining the number of female and male spiders that will be characterized as individuals in the search space. The number of females N_f is randomly selected within the range of 65–90% of the entire population N. Therefore, N_f is calculated by the following equation:

$$N_f = \text{floor}[(0.9 - \text{rand} \cdot 0.25) \cdot N] \tag{2.1}$$

where rand is a random number between [0,1] whereas $\text{floor}(\cdot)$ maps a real number to an integer number. The number of male spiders N_m is computed as the complement between N and N_f. It is calculated as follows:

$$N_m = N - N_f \tag{2.2}$$

Therefore, the complete population $\mathbf{S}$, composed by N elements, is divided in two sub-groups $\mathbf{F}$ and $\mathbf{M}$. The Group $\mathbf{F}$ assembles the set of female individuals $\mathbf{F} = \{\mathbf{f}_1, \mathbf{f}_2, \ldots, \mathbf{f}_{N_f}\}$ whereas $\mathbf{M}$ groups the male members ($\mathbf{M} = \{\mathbf{m}_1, \mathbf{m}_2, \ldots, \mathbf{m}_{N_m}\}$), where $\mathbf{S} = \mathbf{F} \cup \mathbf{M}$ ($\mathbf{S} = \{\mathbf{s}_1, \mathbf{s}_2, \ldots, \mathbf{s}_N\}$), such that $\mathbf{S} = \{\mathbf{s}_1 = \mathbf{f}_1, \mathbf{s}_2 = \mathbf{f}_2, \ldots, \mathbf{s}_{N_f} = \mathbf{f}_{N_f}, \mathbf{s}_{N_f+1} = \mathbf{m}_1, \mathbf{s}_{N_f+2} = \mathbf{m}_2, \ldots, \mathbf{s}_N = \mathbf{m}_{N_m}\}$.

2.3.1 Fitness Assignation

In the biological metaphor, the spider size is the characteristic that evaluates the individual capacity to perform better over its assigned tasks. In the approach, every individual (spider) receives a weight w_i which represents the solution quality that corresponds to the spider i (irrespective of gender) of the population $\mathbf{S}$. In order to calculate the weight of every spider the next equation is used:

$$w_i = \frac{J(\mathbf{s}_i) - worst_{\mathbf{S}}}{best_{\mathbf{S}} - worst_{\mathbf{S}}} \tag{2.3}$$

where $J(\mathbf{s}_i)$ is the fitness value obtained by the evaluation of the spider position $\mathbf{s}_i$ with regard to the objective function $J(\cdot)$. The values $worst_{\mathbf{S}}$ and $best_{\mathbf{S}}$ are defined as follows (considering a maximization problem):

$$best_{\mathbf{S}} = \max_{k \in \{1,2,\ldots,N\}} ((J(\mathbf{s}_k)) \text{ and } worst_{\mathbf{S}} = \min_{k \in \{1,2,\ldots,N\}} ((J(\mathbf{s}_k)) \tag{2.4}$$

2.3.2 Modeling of the Vibrations Through the Communal Web

The communal web is used as a mechanism to transmit information among the colony members. This information is encoded as small vibrations that are critical for the collective coordination of all individuals in the population. The vibrations depend on the weight and distance of the spider which has generated them. Since the distance is relative to the individual that provokes the vibrations and the member who detects them, members located near to the individual that provokes the vibrations, perceive stronger vibrations in comparison with members located in distant positions. In order to reproduce this process, the vibrations perceived by the individual i as a result of the information transmitted by the member j are modeled according to the following equation:

$$Vib_{i,j} = w_j \cdot e^{-d_{i,j}^2} \tag{2.5}$$

where the $d_{i,j}$ is the Euclidian distance between the spiders i and j, such that $d_{i,j} = \|\mathbf{s}_i - \mathbf{s}_j\|$.

Although it is virtually possible to compute perceived-vibrations by considering any pair of individuals, three special relationships are considered within the SSO approach:

1. Vibrations $Vibc_i$ are perceived by the individual i ($\mathbf{s}_i$) as a result of the information transmitted by the member c ($\mathbf{s}_c$) who is an individual that has two important characteristics: it is the nearest member to i and possesses a higher weight in comparison to $i(w_c > w_i)$.

$$Vibc_i = w_c \cdot e^{-d_{i,c}^2} \tag{2.6}$$

2. The vibrations $Vibb_i$ perceived by the individual i as a result of the information transmitted by the member b ($\mathbf{s}_b$), with b being the individual holding the best weight (best fitness value) of the entire population $\mathbf{S}$, such that $w_b = \max_{k \in \{1,2,\ldots,N\}} (w_k)$.

$$Vibb_i = w_b \cdot e^{-d_{i,b}^2} \tag{2.7}$$

3. The vibrations $Vibf_i$ perceived by the individual i ($\mathbf{s}_i$) as a result of the information transmitted by the member f ($\mathbf{s}_f$), with f being the nearest female individual to i.

$$Vibf_i = w_f \cdot e^{-d_{i,f}^2} \tag{2.8}$$

Figure 2.1 shows the configuration of each special relationship: (a) $Vibc_i$, (b) $Vibb_i$ and (c) $Vibf_i$.

2.3.3 Initializing the Population

Like other evolutionary algorithms, the SSO is an iterative process whose first step is to randomly initialize the entire population (female and male). The algorithm begins by initializing the set $\mathbf{S}$ of N spider positions. Each spider position, $\mathbf{f}_i$ or $\mathbf{m}_i$, is a n-dimensional vector containing the parameter values to be optimized. Such values are randomly and uniformly distributed between the pre-specified lower initial parameter bound p_j^{low} and the upper initial parameter bound p_j^{high}, just as it described by the following expressions:

$$\begin{array}{ll} f_{i,j}^0 = p_j^{low} + \text{rand}(0,1) \cdot \left(p_j^{high} - p_j^{low}\right) & m_{k,j}^0 = p_j^{low} + \text{rand}(0,1) \cdot \left(p_j^{high} - p_j^{low}\right) \\ i = 1,2,\ldots, N_f; j = 1,2,\ldots,n & k = 1,2,\ldots, N_m; j = 1,2,\ldots,n \end{array} \tag{2.9}$$

where j, i and k are the parameter and individual indexes respectively whereas zero signals the initial population. The function rand(0,1) generates a random number between 0 and 1. Hence, $f_{i,j}$ is the j-th parameter of the i-th female spider position.

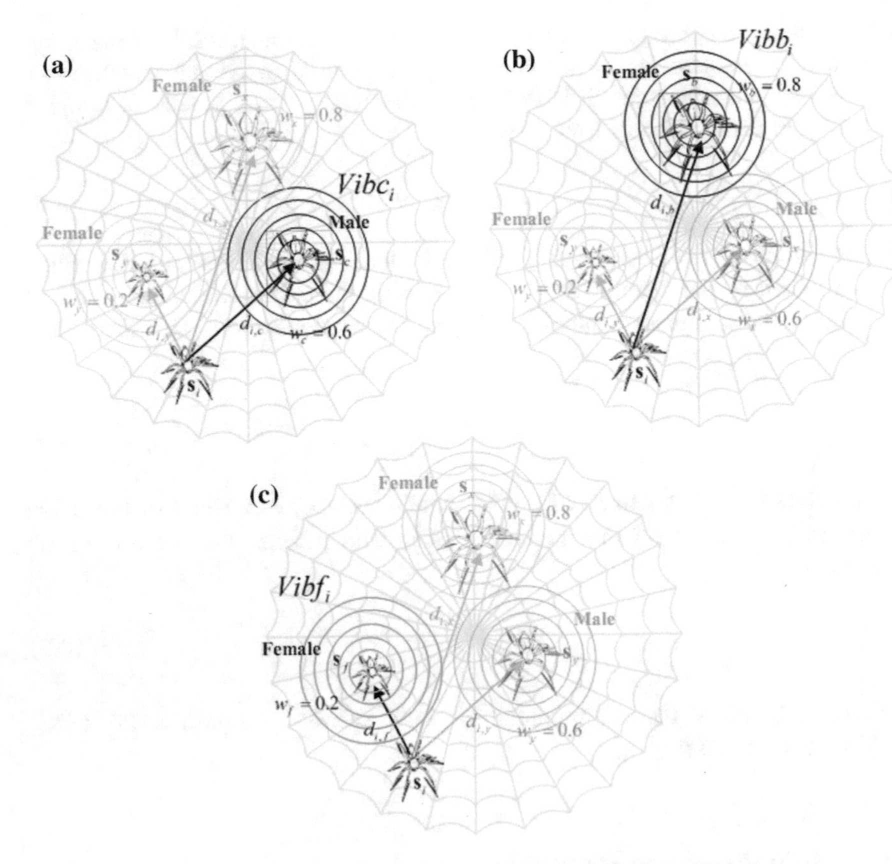

Fig. 2.1 Configuration of each special relation: **a** $Vibc_i$, **b** $Vibb_i$ and **c** $Vibf_i$

2.3.4 Cooperative Operators

Female cooperative operator

Social-spiders perform cooperative interaction over other colony members. The way in which this behavior takes place depends on the spider gender. Female spiders present an attraction or dislike over others irrespective of gender. For a particular female spider, such attraction or dislike is commonly developed over other spiders according to their vibrations which are emitted over the communal web. Since vibrations depend on the weight and distance of the members which have originated them, strong vibrations are produced either by big spiders or other neighboring members lying nearby the individual which is perceiving them. The final decision of attraction or dislike over a determined member is taken considering an internal state which is influenced by several factors such as reproduction cycle, curiosity and other random phenomena.

In order to emulate the cooperative behavior of the female spider, a new operator is defined. The operator considers the position change of the female spider i at each iteration. Such position change, which can be of attraction or repulsion, is computed as a combination of three different elements. The first one involves the change in regard to the nearest member to i that holds a higher weight and produces the vibration $Vibc_i$. The second one considers the change regarding the best individual of the entire population $\mathbf{S}$ who produces the vibration $Vibb_i$. Finally, the third one incorporates a random movement.

Since the final movement of attraction or repulsion depends on several random phenomena, the selection is modeled as a stochastic decision. For this operation, a uniform random number r_m is generated within the range [0,1]. If r_m is smaller than a threshold PF, an attraction movement is generated; otherwise, a repulsion movement is produced. Therefore, such operator can be modeled as follows:

$$\mathbf{f}_i^{k+1} = \begin{cases} \mathbf{f}_i^k + \alpha \cdot Vibc_i \cdot \left(\mathbf{s}_c - \mathbf{f}_i^k\right) + \beta \cdot Vibb_i \cdot \left(\mathbf{s}_b - \mathbf{f}_i^k\right) + \delta \cdot \left(\text{rand} - \frac{1}{2}\right) & \text{with probability } PF \\ \mathbf{f}_i^k - \alpha \cdot Vibc_i \cdot \left(\mathbf{s}_c - \mathbf{f}_i^k\right) - \beta \cdot Vibb_i \cdot \left(\mathbf{s}_b - \mathbf{f}_i^k\right) + \delta \cdot \left(\text{rand} - \frac{1}{2}\right) & \text{with probability } 1 - PF \end{cases} \quad (2.10)$$

where α, β, δ and rand are random numbers between [0,1] whereas k represents the iteration number. The individual $\mathbf{s}_c$ and $\mathbf{s}_b$ represent the nearest member to i that holds a higher weight and the best individual of the entire population $\mathbf{S}$, respectively.

Under this operation, each particle presents a movement which combines the past position that holds the attraction or repulsion vector over the local best element $\mathbf{s}_c$ and the global best individual $\mathbf{s}_b$ seen so-far. This particular type of interaction avoids the quick concentration of particles at only one point and encourages each particle to search around the local candidate region within its neighborhood ($\mathbf{s}_c$), rather than interacting to a particle ($\mathbf{s}_b$) in a distant region of the domain. The use of this scheme has two advantages. First, it prevents the particles from moving towards the global best position, making the algorithm less susceptible to premature convergence. Second, it encourages particles to explore their own neighborhood thoroughly before converging towards the global best position. Therefore, it provides the algorithm with global search ability and enhances the exploitative behavior of the approach.

Male cooperative operator

According to the biological behavior of the social-spider, male population is divided into two classes: dominant and non-dominant male spiders. Dominant male spiders have better fitness characteristics (usually regarding the size) in comparison to non-dominant. Dominant males are attracted to the closest female spider in the communal web. In contrast, non-dominant male spiders tend to concentrate in the center of the male population as a strategy to take advantage of resources that are wasted by dominant males.

For emulating such cooperative behavior, the male members are divided into two different groups (dominant members **D** and non-dominant members **ND**) according to their position with regard to the median member. Male members, with a weight value above the median value within the male population, are considered the dominant individuals **D**. On the other hand, those under the median value are labeled as non-dominant **ND** males. In order to implement such computation, the male population **M** ($\mathbf{M} = \{\mathbf{m}_1, \mathbf{m}_2, \ldots, \mathbf{m}_{N_m}\}$) is arranged according to their weight value in decreasing order. Thus, the individual whose weight w_{N_f+m} is located in the middle is considered the median male member. Since indexes of the male population **M** in regard to the entire population **S** are increased by the number of female members N_f, the median weight is indexed by $N_f + m$. According to this, change of positions for the male spider can be modeled as follows:

$$\mathbf{m}_i^{k+1} = \begin{cases} \mathbf{m}_i^k + \alpha \cdot Vibf_i \cdot \left(\mathbf{s}_f - \mathbf{m}_i^k\right) + \delta \cdot \left(\text{rand} - \frac{1}{2}\right) & \text{if } w_{N_f+i} > w_{N_f+m} \\ \mathbf{m}_i^k + \alpha \cdot \left(\frac{\sum_{h=1}^{N_m} \mathbf{m}_h^k \cdot w_{N_f+h}}{\sum_{h=1}^{N_m} w_{N_f+h}} - \mathbf{m}_i^k\right) & \text{if } w_{N_f+i} \le w_{N_f+m} \end{cases}, \tag{2.11}$$

where the individual $\mathbf{s}_f$ represents the nearest female individual to the male member i whereas $\left(\sum_{h=1}^{N_m} \mathbf{m}_h^k \cdot w_{N_f+h} / \sum_{h=1}^{N_m} w_{N_f+h}\right)$ correspond to the weighted mean of the male population **M**.

By using this operator, two different behaviors are produced. First, the set **D** of particles is attracted to others in order to provoke mating. Such behavior allows incorporating diversity into the population. Second, the set **ND** of particles is attracted to the weighted mean of the male population **M**. This fact is used to partially control the search process according to the average performance of a sub-group of the population. Such mechanism acts as a filter which avoids that very good individuals or extremely bad individuals influence the search process.

2.3.5 Mating Operator

Mating in a social-spider colony is performed by dominant males and the female members. Under such circumstances, when a dominant male $\mathbf{m}_g$ spider ($g \in \mathbf{D}$) locates a set $\mathbf{E}^g$ of female members within a specific range r (range of mating), it mates, forming a new brood $\mathbf{s}_{new}$ which is generated considering all the elements of the set $\mathbf{T}^g$ that, in turn, has been generated by the union $\mathbf{E}^g \cup \mathbf{m}_g$. It is important to emphasize that if the set $\mathbf{E}^g$ is empty, the mating operation is canceled. The range r is defined as a radius which depends on the size of the search space. Such radius r is computed according to the following model:

$$r = \frac{\sum_{j=1}^{n} \left(p_j^{high} - p_j^{low}\right)}{2 \cdot n} \tag{2.12}$$

In the mating process, the weight of each involved spider (elements of $\mathbf{T}^g$) defines the probability of influence for each individual into the new brood. The spiders holding a heavier weight are more likely to influence the new product, while elements with lighter weight have a lower probability. The influence probability Ps_i of each member is assigned by the roulette method, which is defined as follows:

$$Ps_i = \frac{w_i}{\sum_{j \in \mathbf{T}^k} w_j}, \tag{2.13}$$

where $i \in \mathbf{T}^g$.

Once the new spider is formed, it is compared to the new spider candidate $\mathbf{s}_{new}$ holding the worst spider $\mathbf{s}_{wo}$ of the colony, according to their weight values (where $w_{wo} = \min_{l \in \{1,2,\ldots,N\}}(w_l)$). If the new spider is better than the worst spider, the worst spider is replaced by the new one. Otherwise, the new spider is discarded and the population does not suffer changes. In case of replacement, the new spider assumes the gender and index from the replaced spider. Such fact assures that the entire population $\mathbf{S}$ maintains the original rate between female and male members.

In order to demonstrate the mating operation, Fig. 2.2a illustrates a simple optimization problem. As an example, it is assumed a population $\mathbf{S}$ of eight different

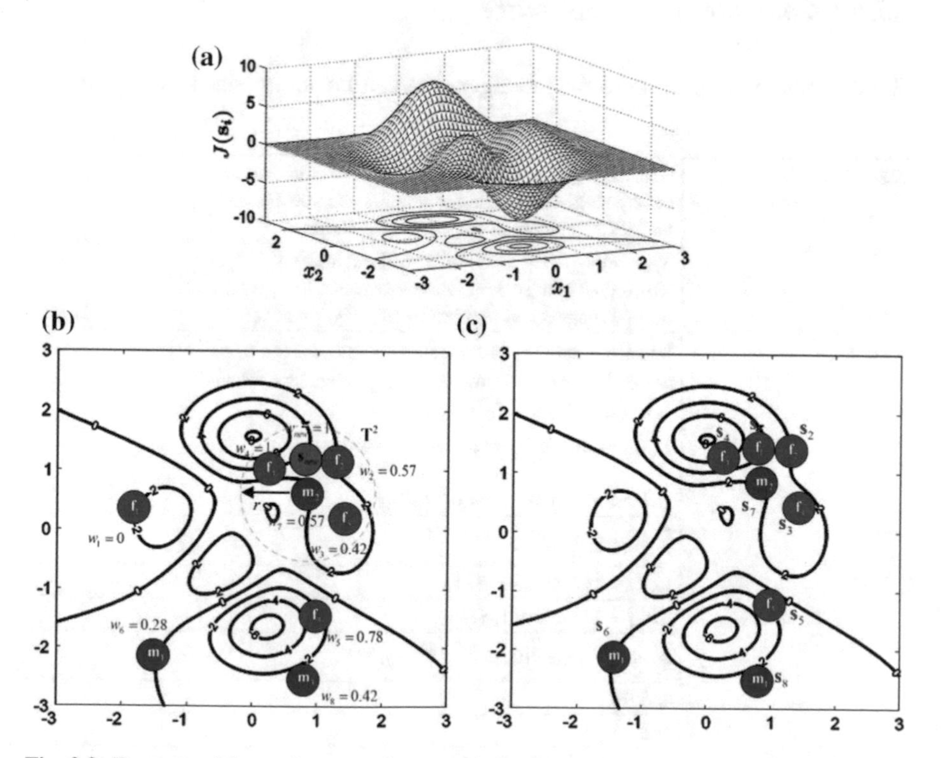

Fig. 2.2 Example of the mating operation: **a** optimization problem, **b** initial configuration before mating and **c** configuration after the mating operation

2-dimensional members ($N = 8$), five females $(N_f = 5)$ and three males $(N_m = 3)$. Figure 2.2b shows the initial configuration of the proposed example with three different female members $\mathbf{f}_2(\mathbf{s}_2)$, $\mathbf{f}_3(\mathbf{s}_3)$ and $\mathbf{f}_4(\mathbf{s}_4)$ constituting the set $\mathbf{E}^2$ which is located inside of the influence range r of a dominant male $\mathbf{m}_2(\mathbf{s}_7)$. Then, the new candidate spider $\mathbf{s}_{new}$ is generated from the elements $\mathbf{f}_2$,$\mathbf{f}_3$, $\mathbf{f}_4$ and $\mathbf{m}_2$ which constitute the set $\mathbf{T}^2$. Therefore, the value of the first decision variable $s_{new,1}$ for the new spider is chosen by means of the roulette mechanism considering the values already existing from the set $\{f_{2,1}, f_{3,1}, f_{4,1}, m_{2,1}\}$. The value of the second decision variable $s_{new,2}$ is also chosen in the same manner. Table 2.1 shows the data for constructing the new spider through the roulette method. Once the new spider $\mathbf{s}_{new}$ is formed, its weight w_{new} is calculated. As $\mathbf{s}_{new}$ is better than the worst member $\mathbf{f}_1$ that is present in the population $\mathbf{S}$, $\mathbf{f}_1$ is replaced by $\mathbf{s}_{new}$. Therefore, $\mathbf{s}_{new}$ assumes the same gender and index from $\mathbf{f}_1$. Figure 2.2c shows the configuration of $\mathbf{S}$ after the mating process.

Under this operation, new generated particles locally exploit the search space inside the mating range in order to find better individuals.

2.3.6 *Computational Procedure*

The computational procedure for the algorithm can be summarized as follows:>

Step 1:	Considering N as the total number of n-dimensional colony members, define the number of male N_m and females N_f spiders in the entire population $\mathbf{S}$
	$N_f = \text{floor}[(0.9 - \text{rand} \cdot 0.25) \cdot N]$ and $N_m = N - N_f$, where rand is a random number between [0,1] whereas floor$(\cdot)$ maps a real number to an integer number
Step 2:	Initialize randomly the female $(\mathbf{F} = \{\mathbf{f}_1, \mathbf{f}_2, \ldots, \mathbf{f}_{N_f}\})$ and male $(\mathbf{M} = \{\mathbf{m}_1, \mathbf{m}_2, \ldots, \mathbf{m}_{N_m}\})$ members (where $\mathbf{S} = \{\mathbf{s}_1 = \mathbf{f}_1, \mathbf{s}_2 = \mathbf{f}_2, \ldots, \mathbf{s}_{N_f} = \mathbf{f}_{N_f},$ $\mathbf{s}_{N_f+1} = \mathbf{m}_1, \mathbf{s}_{N_f+2} = \mathbf{m}_2, \ldots, \mathbf{s}_N = \mathbf{m}_{N_m}\}$ and calculate the radius of mating
	$r = \frac{\sum_{j=1}^{n}\left(p_j^{high} - p_j^{low}\right)}{2 \cdot n}$
	for (i = 1; $i < N_f + 1$; i++)
	for (j = 1; j < n + 1; j++)
	$f_{i,j}^0 = p_j^{low} + \text{rand}(0,1) \cdot \left(p_j^{high} - p_j^{low}\right)$
	end for
	end for
	for (k = 1; $k < N_m + 1$; k++)
	for (j = 1; j < n + 1; j++)

(continued)

(continued)

	$m_{k,j}^{0} = p_j^{low} + \text{rand} \cdot \left(p_j^{high} - p_j^{low}\right)$
	end for
	end for
Step 3:	Calculate the weight of every spider of **S** (Sect. 2.3.1)
	for (i = 1, $i < N + 1$; i++)
	$w_i = \frac{J(\mathbf{s}_i) - worst_{\mathbf{S}}}{best_{\mathbf{S}} - worst_{\mathbf{S}}}$ where $best_{\mathbf{S}} = \max_{k \in \{1,2,\ldots,N\}} (J(\mathbf{s}_k))$ and $worst_{\mathbf{S}} = \min_{k \in \{1,2,\ldots,N\}} (J(\mathbf{s}_k))$
	end for
Step 4:	Move female spiders according to the female cooperative operator (Sect. 2.3.4)
	for (i = 1; $i < N_f + 1$; i++)
	Calculate $Vibc_i$ and $Vibb_i$ (Sect. 2.3.2)
	If ($r_m < PF$); where $r_m \in \text{rand}(0, 1)$
	$\mathbf{f}_i^{k+1} = \mathbf{f}_i^k + \alpha \cdot Vibc_i \cdot \left(\mathbf{s}_c - \mathbf{f}_i^k\right) + \beta \cdot Vibb_i$ $\cdot \left(\mathbf{s}_b - \mathbf{f}_i^k\right) + \delta \cdot \left(\text{rand} - \frac{1}{2}\right)$
	else if
	$\mathbf{f}_i^{k+1} = \mathbf{f}_i^k - \alpha \cdot Vibc_i \cdot \left(\mathbf{s}_c - \mathbf{f}_i^k\right) - \beta \cdot Vibb_i$ $\cdot \left(\mathbf{s}_b - \mathbf{f}_i^k\right) + \delta \cdot \left(\text{rand} - \frac{1}{2}\right)$
	end if
	end for
Step 5:	Move the male spiders according to the male cooperative operator (Sect. 2.3.4)
	Find the median male individual $\left(w_{N_f+m}\right)$ from **M**
	for (i = 1; $i < N_m + 1$; i++)
	Calculate $Vibf_i$ (Sect. 2.3.2)
	If $\left(w_{N_f+i} > w_{N_f+m}\right)$
	$\mathbf{m}_i^{k+1} = \mathbf{m}_i^k + \alpha \cdot Vibf_i \cdot \left(\mathbf{s}_f - \mathbf{m}_i^k\right) + \delta \cdot \left(\text{rand} - \frac{1}{2}\right)$
	Else if
	$\mathbf{m}_i^{k+1} = \mathbf{m}_i^k + \alpha \cdot \left(\frac{\sum_{h=1}^{N_m} \mathbf{m}_h^k \cdot w_{N_f+h}}{\sum_{h=1}^{N_m} w_{N_f+h}} - \mathbf{m}_i^k\right)$
	end if
	end for
Step 6:	Perform the mating operation (Sect. 2.3.5)
	for (i = 1; $i < N_m + 1$; i++)
	If ($\mathbf{m}_i \in \mathbf{D}$)
	Find $\mathbf{E}^i$
	If ($\mathbf{E}^i$ is not empty)
	Form $\mathbf{s}_{new}$ using the roulette method

(continued)

(continued)

	If $(w_{new} > w_{wo})$
	$\mathbf{s}_{wo} = \mathbf{s}_{new}$
	end if
	end if
	end if
	end for
Step 7:	If the stop criteria is met, the process is finished; otherwise, go back to Step 3

Table 2.1 Data for constructing the new spider $\mathbf{s}_{new}$ through the roulette method

Spider		Position	w_i	Ps_i	Roulette
$\mathbf{s}_1$	$\mathbf{f}_1$	(–1.9,0.3)	0.00	–	25% $\mathbf{f}_2$ $\mathbf{f}_4$ 41% $\mathbf{f}_3$ 17% $\mathbf{m}_3$ 17%
$\mathbf{s}_2$	$\mathbf{f}_2$	(1.4,1.1)	0.57	0.22	
$\mathbf{s}_3$	$\mathbf{f}_3$	(1.5,0.2)	0.42	0.16	
$\mathbf{s}_4$	$\mathbf{f}_4$	(0.4,1.0)	1.00	0.39	
$\mathbf{s}_5$	$\mathbf{f}_5$	(1.0,–1.5)	0.78	–	
$\mathbf{s}_6$	$\mathbf{m}_1$	(–1.3,–1.9)	0.28	–	
$\mathbf{s}_7$	$\mathbf{m}_2$	(0.9,0.7)	0.57	0.22	
$\mathbf{s}_8$	$\mathbf{m}_3$	(0.8,–2.6)	0.42	–	
$\mathbf{s}_{new}$		(0.9,1.1)	1.00	–	

2.3.7 Discussion About the SSO Algorithm

Evolutionary algorithms (EA) have been widely employed for solving complex optimization problems. These methods are found to be more powerful than conventional methods based on formal logics or mathematical programming [32]. In an EA algorithm, search agents have to decide whether to explore unknown search positions or to exploit already tested positions in order to improve their solution quality. Pure exploration degrades the precision of the evolutionary process but increases its capacity to find new potential solutions. On the other hand, pure exploitation allows refining existent solutions but adversely drives the process to local optimal solutions. Therefore, the ability of an EA to find a global optimal solutions depends on its capacity to find a good balance between the exploitation of found-so-far elements and the exploration of the search space [33]. So far, the exploration–exploitation dilemma has been an unsolved issue within the framework of evolutionary algorithms.

EA defines individuals with the same property, performing virtually the same behavior. Under these circumstances, algorithms waste the possibility to add new and selective operators as a result of considering individuals with different characteristics. These operators could incorporate computational mechanisms to improve several important algorithm characteristics such as population diversity or searching capacities.

On the other hand, PSO and ABC are the most popular metaheuristic algorithms for solving complex optimization problems. However, they present serious flaws such as premature convergence and difficulty to overcome local minima [10, 11]. Such problems arise from operators that modify individual positions. In such algorithms, the position of each agent in the next iteration is updated yielding an attraction towards the position of the best particle seen so-far (in case of PSO) or any other randomly chosen individual (in case of ABC). Such behaviors produce that the entire population concentrates around the best particle or diverges without control as the algorithm evolves, either favoring the premature convergence or damaging the exploration–exploitation balance [12, 13].

Different to other EA, at SSO each individual is modeled considering the gender. Such fact allows incorporating computational mechanisms to avoid critical flaws such as premature convergence and incorrect exploration–exploitation balance commonly present in both, the PSO and the ABC algorithm. From an optimization point of view, the use of the social-spider behavior as a metaphor introduces interesting concepts in EA: the fact of dividing the entire population into different search-agent categories and the employment of specialized operators that are applied selectively to each of them. By using this framework, it is possible to improve the balance between exploitation and exploration, yet preserving the same population, i.e. individuals who have achieved efficient exploration (female spiders) and individuals that verify extensive exploitation (male spiders). Furthermore, the social-spider behavior mechanism introduces an interesting computational scheme with three important particularities: first, individuals are separately processed according to their characteristics. Second, operators share the same communication mechanism allowing the employment of important information of the evolutionary process to modify the influence of each operator. Third, although operators modify the position of only an individual type, they use global information (positions of all individual types) in order to perform such modification. Figure 2.3 presents a schematic representation of the algorithm-data-flow. According to Fig. 2.3, the female cooperative and male cooperative operators process only female or male individuals, respectively. However, the mating operator modifies both individual types.

2.4 Experimental Results

A comprehensive set of 19 functions, which have been collected from Refs. [34–40], has been used to test the performance of the SSO approach. Table 2.4 in the Appendix presents the benchmark functions used in our experimental study. In the table, n indicates the function dimension, $f(\mathbf{x}^*)$ the optimum value of the function, $\mathbf{x}^*$

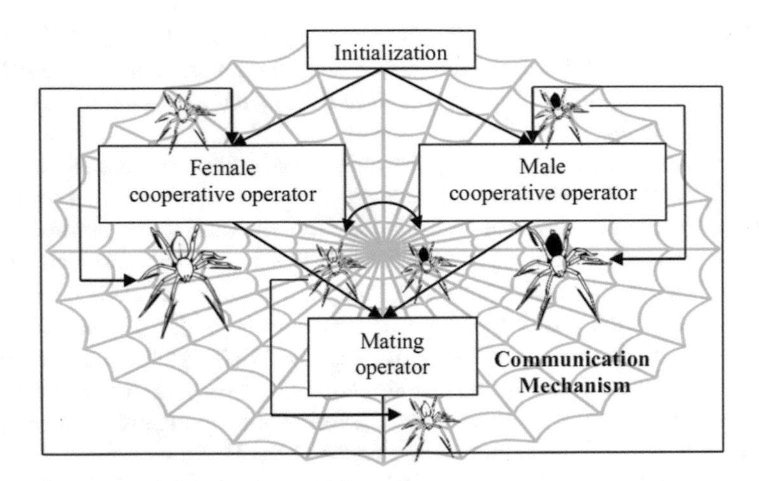

Fig. 2.3 Schematic representation of the SSO algorithm-data-flow

the optimum position and S the search space (subset of R^n). A detailed description of each function is given in the Appendix.

2.4.1 *Performance Comparison to Other Metaheuristic Algorithms*

We have applied the SSO algorithm to 19 functions whose results have been compared to those produced by the Particle Swarm Optimization (PSO) method [3] and the Artificial Bee Colony (ABC) algorithm [4]. These are considered as the most popular metaheuristic algorithms for many optimization applications. In all comparisons, the population has been set to 50 individuals. The maximum iteration number for all functions has been set to 1000. Such stop criterion has been selected to maintain compatibility to similar works reported in the literature [41, 42].

The parameter setting for each algorithm in the comparison is described as follows:

1. PSO: The parameters are set to $c_1 = 2$ and $c_2 = 2$; besides, the weight factor decreases linearly from 0.9 to 0.2 [3].
2. ABC: The algorithm has been implemented using the guidelines provided by its own reference [4], using the parameter *limit* = 100.
3. SSO: Once it has been determined experimentally, the parameter *PF* has been set to 0.7. It is kept for all experiments in this section.

The experiment compares the SSO to other algorithms such as PSO and ABC. The results for 30 runs are reported in Table 2.2 considering the following performance indexes: the Average Best-so-far (AB) solution, the Median Best-so-far (MB) and the Standard Deviation (SD) of best-so-far solution. The best outcome for

Table 2.2 Minimization results of benchmark functions of Table 2.4 with n = 30

		SSO	ABC	PSO
$f_1(x)$	**AB**	**1.96E−03**	2.90E−03	1.00E+03
	MB	2.81E−03	1.50E−03	**2.08E−09**
	SD	**9.96E−04**	1.44E−03	3.05E+03
$f_2(x)$	**AB**	**1.37E−02**	1.35E−01	5.17E+01
	MB	**1.34E−02**	1.05E−01	5.00E+01
	SD	**3.11E−03**	8.01E−02	2.02E+01
$f_3(x)$	**AB**	**4.27E−02**	1.13E+00	8.63E+04
	MB	**3.49E−02**	6.11E−01	8.00E+04
	SD	**3.11E−02**	1.57E+00	5.56E+04
$f_4(x)$	**AB**	**5.40E−02**	5.82E+01	1.47E+01
	MB	**5.43E−02**	5.92E+01	1.51E+01
	SD	**1.01E−02**	7.02E+00	3.13E+00
$f_5(x)$	**AB**	**1.14E+02**	1.38E+02	3.34E+04
	MB	**5.86E+01**	1.32E+02	4.03E+02
	SD	**3.90E+01**	1.55E+02	4.38E+04
$f_6(x)$	**AB**	**2.68E−03**	4.06E−03	1.00E+03
	MB	2.68E−03	3.74E−03	**1.66E−09**
	SD	**6.05E−04**	2.98E−03	3.06E+03
$f_7(x)$	**AB**	**1.20E+01**	1.21E+01	1.50E+01
	MB	**1.20E+01**	1.23E+01	1.37E+01
	SD	**5.76E−01**	9.00E−01	4.75E+00
$f_8(x)$	**AB**	**2.14E+00**	3.60E+00	3.12E+04
	MB	**3.64E+00**	8.04E−01	2.08E+02
	SD	**1.26E+00**	3.54E+00	5.74E+04
$f_9(x)$	**AB**	**6.92E−05**	1.44E−04	2.47E+00
	MB	**6.80E−05**	8.09E−05	9.09E−01
	SD	**4.02E−05**	1.69E−04	3.27E+00
$f_{10}(x)$	**AB**	**4.44E−04**	1.10E−01	6.93E+02
	MB	**4.05E−04**	4.97E−02	5.50E+02
	SD	**2.90E−04**	1.98E−01	6.48E+02
$f_{11}(x)$	**AB**	**6.81E+01**	3.12E+02	4.11E+02
	MB	**6.12E+01**	3.13E+02	4.31E+02
	SD	**3.00E+01**	4.31E+01	1.56E+02
$f_{12}(x)$	**AB**	**5.39E−05**	1.18E−04	4.27E+07
	MB	**5.40E−05**	1.05E−04	1.04E−01
	SD	**1.84E−05**	8.88E−05	9.70E+07
$f_{13}(x)$	**AB**	**1.76E−03**	1.87E−03	5.74E−01
	MB	**1.12E−03**	1.69E−03	1.08E−05
	SD	**6.75E−04**	1.47E−03	2.36E+00

(continued)

Table 2.2 (continued)

		SSO	ABC	PSO
$f_{14}(x)$	**AB**	**−9.36E+02**	−9.69E+02	−9.63E+02
	MB	**−9.36E+02**	−9.60E+02	−9.92E+02
	SD	**1.61E+01**	6.55E+01	6.66E+01
$f_{15}(x)$	**AB**	**8.59E+00**	2.64E+01	1.35E+02
	MB	**8.78E+00**	2.24E+01	1.36E+02
	SD	**1.11E+00**	1.06E+01	3.73E+01
$f_{16}(x)$	**AB**	**1.36E−02**	6.53E−01	1.14E+01
	MB	**1.39E−02**	6.39E−01	1.43E+01
	SD	**2.36E−03**	3.09E−01	8.86E+00
$f_{17}(x)$	**AB**	**3.29E−03**	5.22E−02	1.20E+01
	MB	**3.21E−03**	4.60E−02	1.35E−02
	SD	**5.49E−04**	3.42E−02	3.12E+01
$f_{18}(x)$	**AB**	**1.87E+00**	2.13E+00	1.26E+03
	MB	**1.61E+00**	2.14E+00	5.67E+02
	SD	**1.20E+00**	1.22E+00	1.12E+03
$f_{19}(x)$	**AB**	**2.74E−01**	4.14E+00	1.53E+00
	MB	**3.00E−01**	4.10E+00	5.50E−01
	SD	**5.17E−02**	4.69E−01	2.94E+00

Maximum number of iterations = 1000
Bold data represents the best values

each function is boldfaced. According to this table, SSO delivers better results than PSO and ABC for all functions. In particular, the test remarks the largest difference in performance which is directly related to a better trade-off between exploration and exploitation. Figure 2.4 presents the evolution curves for PSO, ABC and the SSO algorithm considering as examples the functions $f_1, f_3, f_5, f_{10}, f_{15}$ and f_{19} from the experimental set. Among them, the rate of convergence of SSO is the fastest, which finds the best solution in less of 400 iterations on average while the other three algorithms need much more iterations. A non-parametric statistical significance proof known as the Wilcoxon's rank sum test for independent samples [43, 44] has been conducted over the "average best-so-far" (AB) data of Table 2.2, with an 5% significance level. Table 2.3 reports the p-values produced by Wilcoxon's test for the pair-wise comparison of the "average best so-far" of two groups. Such groups are constituted by SSO versus PSO and SSO versus ABC. As a null hypothesis, it is assumed that there is no significant difference between mean values of the two algorithms. The alternative hypothesis considers a significant difference between the "average best-so-far" values of both approaches. All p-values reported in Table 2.3 are less than 0.05 (5% significance level) which is a strong evidence against the null hypothesis. Therefore, such evidence indicates that SSO results are statistically significant and it has not occurred by coincidence (i.e. due to common noise contained in the process).

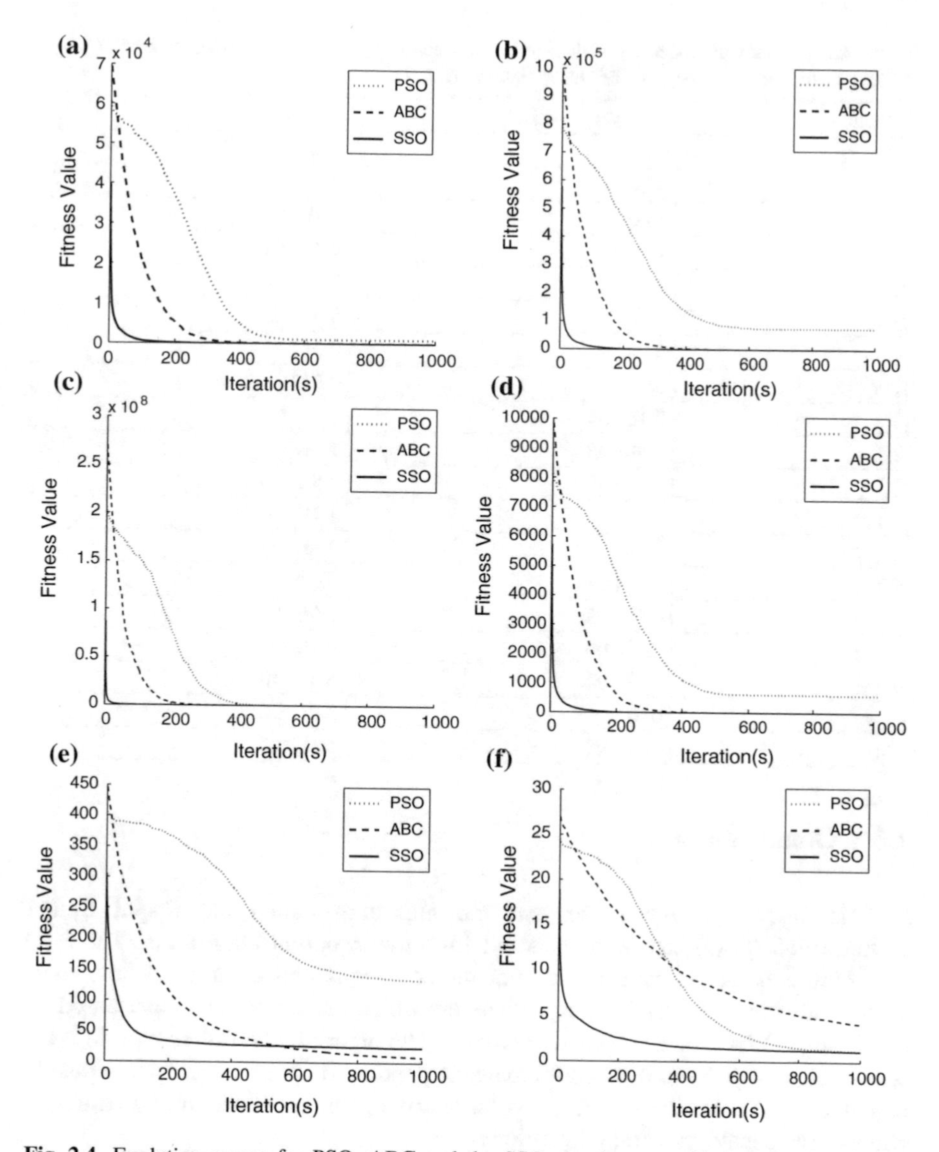

Fig. 2.4 Evolution curves for PSO, ABC and the SSO algorithm considering as examples the functions **a** f_1, **b** f_3, **c** f_5, **d** f_{10}, **e** f_{15} and **f** f_{19} from the experimental set

Table 2.3 p-values produced by Wilcoxon's test comparing SSO versus ABC and SSO versus PSO, over the "average best-so-far" (AB) values from Table 2.2

Function	SSO versus ABC	SSO versus PSO
$f_1(x)$	0.041	1.8E−05
$f_2(x)$	0.048	0.059
$f_3(x)$	5.4E−04	6.2E−07
$f_4(x)$	1.4E−07	4.7E−05
$f_5(x)$	0.045	7.1E−07
$f_6(x)$	2.3E−04	5.5E−08
$f_7(x)$	0.048	0.011
$f_8(x)$	0.017	0.043
$f_9(x)$	8.1E−04	2.5E−08
$f_{10}(x)$	4.6E−06	1.7E−09
$f_{11}(x)$	9.2E−05	7.8E−06
$f_{12}(x)$	0.022	1.1E−10
$f_{13}(x)$	0.048	2.6E−05
$f_{14}(x)$	0.044	0.049
$f_{15}(x)$	4.5E−05	7.9E−08
$f_{16}(x)$	2.8E−05	4.1E−06
$f_{17}(x)$	7.1E−04	6.2E−10
$f_{18}(x)$	0.013	8.3E−10
$f_{19}(x)$	4.9E−05	5.1E−08

2.5 Conclusions

In this chapter, a novel metaheuristic algorithm called the Social Spider Optimization (SSO) has been proposed for solving optimization tasks. The SSO algorithm is based on the simulation of the cooperative behavior of social-spiders whose individuals emulate a group of spiders which interact to each other based on the biological laws of a cooperative colony. The algorithm considers two different search agents (spiders): male and female. Depending on gender, each individual is conducted by a set of different evolutionary operators which mimic different cooperative behaviors within the colony.

In contrast to most of existent metaheuristic algorithms, the SSO approach models each individual considering two genders. Such fact allows not only to emulate the cooperative behavior of the colony in a realistic way, but also to incorporate computational mechanisms to avoid critical flaws commonly delivered by the popular PSO and ABC algorithms, such as the premature convergence and the incorrect exploration–exploitation balance.

SSO has been experimentally tested considering a suite of 19 benchmark functions. The performance of SSO has been also compared to the following metaheuristic algorithms: the Particle Swarm Optimization method (PSO) [16], and the Artificial Bee Colony (ABC) algorithm [38]. Results have confirmed a acceptable performance of the SSO method in terms of the solution quality of the solution for all tested benchmark functions.

The SSO's remarkable performance is associated with two different reasons: (i) their operators allow a better particle distribution in the search space, increasing the algorithm's ability to find the global optima; and (ii) the division of the population into different individual types, provides the use of different rates between exploration and exploitation during the evolution process.

Appendix: List of Benchmark Functions

See Table 2.4.

Table 2.4 Test functions used in the experimental study

Name	Function	S	Dim	Minimum
Sphere	$f_1(\mathbf{x}) = \sum_{i=1}^{n} x_i^2$	$[-100,100]^n$	$n = 30$	$\mathbf{x}^* = (0,\ldots,0)$; $f(\mathbf{x}^*) = 0$
Schwefel 2.22	$f_2(\mathbf{x}) = \sum_{i=1}^{n} \lvert x_i \rvert + \prod_{i=1}^{n} \lvert x_i \rvert$	$[-10,10]^n$	$n = 30$	$\mathbf{x}^* = (0,\ldots,0)$; $f(\mathbf{x}^*) = 0$
Schwefel 1.2	$f_3(\mathbf{x}) = \sum_{i=1}^{n} \left(\sum_{j=1}^{i} x_j\right)^2$	$[-100,100]^n$	$n = 30$	$\mathbf{x}^* = (0,\ldots,0)$; $f(\mathbf{x}^*) = 0$
F4	$f_4(\mathbf{x}) = 418.9829n + \sum_{i=1}^{n} \left(-x_i \sin\left(\sqrt{\lvert x_i \rvert}\right)\right)$	$[-100,100]^n$	$n = 30$	$\mathbf{x}^* = (0,\ldots,0)$; $f(\mathbf{x}^*) = 0$
Rosenbrock	$f_5(\mathbf{x}) = \sum_{i=1}^{n-1} \left[100(x_{i+1} - x_i^2)^2 + (x_i - 1)^2\right]$	$[-30,30]^n$	$n = 30$	$\mathbf{x}^* = (1,\ldots,1)$; $f(\mathbf{x}^*) = 0$
Step	$f_6(\mathbf{x}) = \sum_{i=1}^{n} (\lfloor x_i + 0.5 \rfloor)^2$	$[-100,100]^n$	$n = 30$	$\mathbf{x}^* = (0,\ldots,0)$; $f(\mathbf{x}^*) = 0$
Quartic	$f_7(\mathbf{x}) = \sum_{i=1}^{n} i x_i^4 + random(0,1)$	$[-1.28,1.28]^n$	$n = 30$	$\mathbf{x}^* = (0,\ldots,0)$; $f(\mathbf{x}^*) = 0$
Dixon and Price	$f_8(\mathbf{x}) = (x_1 - 1)^2 + \sum_{i=1}^{n} i(2x_i^2 - x_{i-1})^2$	$[-10,10]^n$	$n = 30$	$\mathbf{x}^* = (0,\ldots,0)$; $f(\mathbf{x}^*) = 0$
Levy	$f_9(\mathbf{x}) = 0.1\left\{\sin^2(3\pi x_1) + \sum_{i=1}^{n} (x_i - 1)^2 [1 + \sin^2(3\pi x_i + 1)] + (x_n - 1)^2 [1 + \sin^2(2\pi x_n)]\right\} + \sum_{i=1}^{n} u(x_i, 5, 100, 4)$; $u(x_i, a, k, m) = \begin{cases} k(x_i - a)^m & x_i > a \\ 0 & -a < x_i < a \\ k(-x_i - a)^m & x_i < -a \end{cases}$	$[-10,10]^n$	$n = 30$	$\mathbf{x}^* = (1,\ldots,1)$; $f(\mathbf{x}^*) = 0$
Sum of squares	$f_{10}(\mathbf{x}) = \sum_{i=1}^{n} i x_i^2$	$[-10,10]^n$	$n = 30$	$\mathbf{x}^* = (0,\ldots,0)$; $f(\mathbf{x}^*) = 0$
Zakharov	$f_{11}(\mathbf{x}) = \sum_{i=1}^{n} x_i^2 + \left(\sum_{i=1}^{n} 0.5 i x_i\right)^2 + \left(\sum_{i=1}^{n} 0.5 i x_i\right)^4$	$[-5,10]^n$	$n = 30$	$\mathbf{x}^* = (0,\ldots,0)$; $f(\mathbf{x}^*) = 0$

(continued)

Table 2.4 (continued)

Name	Function	S	Dim	Minimum
Penalized	$f_{12}(\mathbf{x}) = \frac{\pi}{n}\left\{10\sin(\pi y_1) + \sum_{i=1}^{n-1}(y_i-1)^2\left[1+10\sin^2(\pi y_{i+1})\right] + (y_n-1)^2\right\} + \sum_{i=1}^{n} u(x_i, 10, 100, 4)$ $y_i = 1 + \frac{(x_i+1)}{4}$ $u(x_i,a,k,m) = \begin{cases} k(x_i-a)^m & x_i > a \\ 0 & -a \le x_i \le a \\ k(-x_i-a)^m & x_i < a \end{cases}$	$[-50,50]^n$	$n = 30$	$\mathbf{x}^* = (0,\ldots,0)$; $f(\mathbf{x}^*) = 0$
Penalized 2	$f_{13}(\mathbf{x}) = 0.1\left\{\sin^2(3\pi x_1) + \sum_{i=1}^{n}(x_i-1)^2\left[1+\sin^2(3\pi x_i+1)\right] + (x_n-1)^2\left[1+\sin^2(2\pi x_n)\right]\right\} + \sum_{i=1}^{n} u(x_i, 5, 100, 4)$ where $u(x_i,a,k,m)$ is the same as Penalized function	$[-50,50]^n$	$n = 30$	$\mathbf{x}^* = (0,\ldots,0)$; $f(\mathbf{x}^*) = 0$
Schwefel	$f_{14}(\mathbf{x}) = \sum_{i=1}^{n} -x_i \sin\left(\sqrt{\lvert x_i \rvert}\right)$	$[-500,500]^n$	$n = 30$	$\mathbf{x}^* = (420,\ldots,420)$; $f(\mathbf{x}^*) = -418.9829 \times n$
Rastrigin	$f_{15}(\mathbf{x}) = \sum_{i=1}^{n}\left[x_i^2 - 10\cos(2\pi x_i) + 10\right]$	$[-5.12,5.12]^n$	$n = 30$	$\mathbf{x}^* = (0,\ldots,0)$; $f(\mathbf{x}^*) = 0$
Ackley	$f_{16}(\mathbf{x}) = -20\exp\left(-0.2\sqrt{\frac{1}{n}\sum_{i=1}^{n} x_i^2}\right) - \exp\left(\frac{1}{n}\sum_{i=1}^{n}\cos(2\pi x_i)\right) + 20 + \exp$	$[-32,32]^n$	$n = 30$	$\mathbf{x}^* = (0,\ldots,0)$; $f(\mathbf{x}^*) = 0$
Griewank	$f_{17}(\mathbf{x}) = \frac{1}{4000}\sum_{i=1}^{n} x_i^2 - \prod_{i=1}^{n}\cos\left(\frac{x_i}{\sqrt{i}}\right) + 1$	$[-600,600]^n$	$n = 30$	$\mathbf{x}^* = (0,\ldots,0)$; $f(\mathbf{x}^*) = 0$
Powelll	$f_{18}(\mathbf{x}) = \sum_{i=1}^{n/k}(x_{4i-3} + 10x_{4i-2})^2 + 5(x_{4i-1} - x_{4i})^2 + (x_{4i-2} - x_{4i-1})^4 + 10(x_{4i-3} - x_{4i})^4$	$[-4,5]^n$	$n = 30$	$\mathbf{x}^* = (0,\ldots,0)$; $f(\mathbf{x}^*) = 0$
Salomon	$f_{19}(\mathbf{x}) = -\cos\left(2\pi\sqrt{\sum_{i=1}^{n} x_i^2}\right) + 0.1\sqrt{\sum_{i=1}^{n} x_i^2} + 1$	$[-100,100]^n$	$n = 30$	$\mathbf{x}^* = (0,\ldots,0)$; $f(\mathbf{x}^*) = 0$

References

1. Bonabeau, E., Dorigo, M., Theraulaz, G.: Swarm Intelligence: From Natural to Artificial Systems. Oxford University Press Inc, New York (1999)
2. Kassabalidis, I., El-Sharkawi, M.A., Marks II, R.J., Arabshahi, P., Gray, A.A.: Swarm intelligence for routing in communication networks. In: Global Telecommunications Conference, GLOBECOM '01, IEEE, vol. 6, pp. 3613–3617 (2001)
3. Kennedy, J., Eberhart, R.: Particle swarm optimization. In: Proceedings of the 1995 IEEE International Conference on Neural Networks, vol. 4, pp. 1942–1948, December 1995
4. Karaboga, D.: An idea based on honey bee swarm for numerical optimization. Technical Report-TR06. Engineering Faculty, Computer Engineering Department, Erciyes University (2005)
5. Passino, K.M.: Biomimicry of bacterial foraging for distributed optimization and control. IEEE Control Syst. Mag. **22**(3), 52–67 (2002)
6. Hossein, A., Hossein-Alavi, A.: Krill herd: a new bio-inspired optimization algorithm. Commun. Nonlinear Sci. Numer. Simul. **17**, 4831–4845 (2012)
7. Yang, X.S: Engineering Optimization: An Introduction with Metaheuristic Applications. Wiley, USA (2010)
8. Rajabioun, R.: Cuckoo Optimization Algorithm. Appl. Soft Comput. **11**, 5508–5518 (2011)
9. Bonabeau, E.: Social insect colonies as complex adaptive systems. Ecosystems **1**, 437–443 (1998)
10. Wang, Y., Li, B., Weise, T., Wang, J., Yuan, B., Tian, Q.: Self-adaptive learning based particle swarm optimization. Inf. Sci. **181**(20), 4515–4538 (2011)
11. Wan-li, X., Mei-qing, A.: An efficient and robust artificial bee colony algorithm for numerical optimization. Comput. Oper. Res. **40**, 1256–1265 (2013)
12. Wang, H., Sun, H., Li, C., Rahnamayan, S., Jeng-shyang, P.: Diversity enhanced particle swarm optimization with neighborhood. Inf. Sci. **223**, 119–135 (2013)
13. Banharnsakun, A., Achalakul, T., Sirinaovakul, B.: The best-so-far selection in artificial bee colony algorithm. Appl. Soft Comput. **11**, 2888–2901 (2011)
14. Gordon, D.: The organization of work in social insect colonies. Complexity **8**(1), 43–46 (2003)
15. Lubin, T.B.: The evolution of sociality in spiders. In Brockmann, H.J. (ed.) Advances in the Study of Behavior, vol. 37, pp. 83–145 (2007)
16. Uetz, G.W.: Colonial web-building spiders: balancing the costs and benefits of group-living. In: Choe, E.J., Crespi, B. (eds.) The Evolution of Social Behavior in Insects and Arachnids, pp. 458–475. Cambridge University Press, Cambridge (1997)
17. Aviles, L.: Sex-ratio bias and possible group selection in the social spider anelosimus eximius. Am. Nat. **128**(1), 1–12 (1986)
18. Burgess, J.W.: Social spacing strategies in spiders. In: Rovner, P.N. (ed.) Spider Communication: Mechanisms and Ecological Significance, pp. 317–351. Princeton University Press, Princeton (1982)
19. Maxence, S.: Social organization of the colonial spider Leucauge sp. in the neotropics: vertical stratification within colonies. J Arachnology **38**, 446–451 (2010)
20. Yip, E.C., Powers, K.S., Avilés, L.: Cooperative capture of large prey solves scaling challenge faced by spider societies. Proc. Nat. Acad. Sci. U.S.A. **105**(33), 11818–11822 (2008)
21. Oster, G., Wilson, E.: Caste and Ecology in the Social Insects. Princeton University Press, Princeton (1978)
22. Hölldobler, B., Wilson, E.O.: Journey to the Ants: A Story of Scientific Exploration. ISBN 0-674-48525-4 (1994)
23. Hölldobler, B., Wilson, E.O.: The Ants. Harvard University Press, USA. ISBN 0-674-04075-9 (1990)

24. Avilés, L.: Causes and consequences of cooperation and permanent-sociality in spiders. In: Choe, B.C. (ed.) The Evolution of Social Behavior in Insects and Arachnids, pp. 476–498. Cambridge University Press, Cambridge (1997)
25. Rayor, E.C.: Do social spiders cooperate in predator defense and foraging without a web? Behav. Ecol. Sociobiol. **65**(10), 1935–1945 (2011)
26. Gove, R., Hayworth, M., Chhetri, M., Rueppell, O.: Division of labour and social insect colony performance in relation to task and mating number under two alternative response threshold models. Insect. Soc. **56**(3), 19–331 (2009)
27. Rypstra, A.L., Prey Size, R.S.: Prey perishability and group foraging in a social spider. Oecologia **86**(1), 25–30 (1991)
28. Pasquet, A.: Cooperation and prey capture efficiency in a social spider, Anelosimus eximius (Araneae, Theridiidae). Ethology **90**, 121–133 (1991)
29. Ulbrich, K., Henschel, J.: Intraspecific competition in a social spider. Ecol. Model. **115**(2–3), 243–251 (1999)
30. Jones, T., Riechert, S.: Patterns of reproductive success associated with social structure and microclimate in a spider system. Anim. Behav. **76**(6), 2011–2019 (2008)
31. Damian, O., Andrade, M., Kasumovic, M.: Dynamic population structure and the evolution of spider mating systems. Adv. Insect Physiol. **41**, 65–114 (2011)
32. Yang, X.-S.: Nature-Inspired Metaheuristic Algorithms. Luniver Press, Beckington (2008)
33. Chen, D.B., Zhao, C.X.: Particle swarm optimization with adaptive population size and its application. Appl. Soft Comput. **9**(1), 39–48 (2009)
34. Storn, R., Price, K.: Differential evolution—a simple and efficient heuristic for global optimization over continuous spaces. J. Global Optim. **11**(4), 341–359 (1995)
35. Yang, E., Barton, N.H., Arslan, T., Erdogan, A.T.: A novel shifting balance theory-based approach to optimization of an energy-constrained modulation scheme for wireless sensor networks. In: Proceedings of the IEEE Congress on Evolutionary Computation, CEC 2008, Hong Kong, China, IEEE, pp. 2749–2756, 1–6 June 2008
36. Duan, X., Wang, G.G., Kang, X., Niu, Q., Naterer, G., Peng, Q.: Performance study of mode-pursuing sampling method. Eng. Optim. **41**(1), 1–21 (2009)
37. Vesterstrom, J., Thomsen, R.: A comparative study of differential evolution, particle swarm optimization, and evolutionary algorithms on numerical benchmark problems. In: Congress on Evolutionary Computation, 2004, CEC 2004, vol. 2, pp. 1980–1987, 19–23 June 2004
38. Mezura-Montes, E., Velázquez-Reyes, J., Coello Coello, C.A.: A comparative study of differential evolution variants for global optimization. In: Proceedings of the 8th Annual Conference on Genetic and Evolutionary Computation (GECCO '06), ACM, New York, USA, pp. 485–492 (2006)
39. Karaboga, D., Akay, B.: A comparative study of artificial bee colony algorithm. App. Math. Comput. **214**(1), 108–132 (2009). ISSN 0096-3003
40. Krishnanand, K.R., Nayak, S.K., Panigrahi, B.K., Rout, P.K.: Comparative study of five bio-inspired evolutionary optimization techniques. In: World Congress on Nature & Biologically Inspired Computing, NaBIC, pp. 1231–1236 (2009)
41. Ying, J., Ke-Cun, Z., Shao-Jian, Q.: A deterministic global optimization algorithm. Appl. Math. Comput. **185**(1), 382–387 (2007)
42. Rashedia, E., Nezamabadi-pour, H., Saryazdi, S.: Filter modeling using gravitational search algorithm. Eng. Appl. Artif. Intell. **24**(1), 117–122 (2011)
43. Wilcoxon, F.: Individual comparisons by ranking methods. Biometrics **1**, 80–83 (1945)
44. Garcia, S., Molina, D., Lozano, M., Herrera, F.: A study on the use of non-parametric tests for analyzing the evolutionary algorithms' behaviour: a case study on the CEC '2005 Special session on real parameter optimization. J Heurist (2008). https://doi.org/10.1007/s10732-008-9080-4

Chapter 3
Calibration of Fractional Fuzzy Controllers by Using the Social-Spider Method

Fuzzy controllers (FCs) based on integer concepts have proved their interesting capacities in several engineering domains. The fact that dynamic processes can be more precisely modeled by using fractional systems has generated a great interest in considering the design of FCs under fractional principles. In the design of fractional FCs, the parameter adjustment operation is converted into a multidimensional optimization task where fractional orders, and controller parameters, are assumed as decision elements. In the design of fractional FCs, the complexity of the optimization problem produces multi-modal error surfaces which are significantly hard to solve. Several metaheuristic algorithms have been successfully used to identify the optimal elements of fractional FCs. But, most of them present a big weakness since they usually get sub-optimal solutions as a result of their improper balance between exploitation and exploration in their search process. This chapter analyses the optimal parameter calibration of fractional FCs. To determine the best elements, the approach employs the Social Spider Optimization (SSO) algorithm, which is based on the simulation of the cooperative operation of social-spiders. In SSO, candidate solutions represent a group of spiders, which interact with each other by considering the biological concepts of the spider colony. Different to most of the metaheuristic algorithms, the approach explicitly avoids the concentration of solutions in the promising positions, eliminating critical defects such as the premature convergence and the deficient balance of exploration–exploitation.

3.1 Introduction

A fractional order model is a system that is characterized by a fractional differential equation containing derivatives of non-integer order. In fractional calculus, the integration and the differentiation operators are generalized into a non-integer order element, where is a fractional number and a and t symbolize the operator limits [1, 2]. Several dynamic systems can be more accurately described and controlled by

E. Cuevas et al., *Advances in Metaheuristics Algorithms: Methods and Applications*, Studies in Computational Intelligence 775,
https://doi.org/10.1007/978-3-319-89309-9_3

fractional models in comparison to integer order schemes. For this reason, in the last decade, the fractional order controllers [3–5] have attracted the interests of several research communities.

A fractional-order controller incorporates an integrator of order and a differentiator of order. The superior performance of such controllers with regard to conventional PIDs has been widely demonstrated in the literature [6].

On the other hand, fuzzy logic [7] provides an alternative method to design controllers through the use of heuristic information. Remarkably, such heuristic information may come from a human-operator who directly manipulates the process. In the fuzzy logic methodology, a human operator defines a set of rules on how to control a process, then incorporated it into a fuzzy controller that emulates the decision-making process of the operator [8].

Fractional fuzzy controllers (FFCs) are the results of the combination of conventional fuzzy controllers and fractional operators. Under such combination, FFCs exhibit better results than conventional FCs for an extensive variety of dynamical systems. This capacity is attributed to the additional flexibility offered by the inclusion of the fractional parameters.

Parameter calibration is an important step to implement applications with FFCs. This procedure is long and time consuming, since it is commonly conducted through trial and error. Therefore, the problem of parameter estimation in FFCs can be handled by evolutionary optimization methods. In general, they have demonstrated to deliver interesting results in terms of accuracy and robustness [5]. In these methods, an individual is represented by a candidate parameter set, which symbolizes the configuration of a determined fractional fuzzy controller. Just as the evolution process unfolds, a set of evolutionary operators is applied in order to produce better individuals. The quality of each candidate solution is evaluated through an objective function whose final result represents the performance of the parameter set in terms of the produced error. Some examples of such approaches being applied to the identification of fractional order systems have involved methods such as Genetic Algorithms (GA) [5], Particle Swarm Optimization (PSO) [9], Harmony Search (HS) [10], Gravitational Search Algorithm (GSA) [11] and Cuckoo Search (CS) [12]. Although such algorithms present interesting results, they have exhibited an important limitation: they frequently obtained sub-optimal solutions as a consequence of the limited balance between exploration and exploitation in their search strategies. Such limitation is associated to the evolutionary operators that have been employed to modify individual positions. In such algorithms, during their operation, the position of each individual for the next iteration is updated, producing an attraction towards the position of the best particle seen so far or towards other promising individuals. Therefore, as the algorithm evolves, such behaviors cause that the entire population concentrate rapidly around the best particles, favoring the premature convergence and damaging the appropriate exploration of the search space [13, 14].

The Social Spider Optimization (SSO) algorithm [15] is a recent evolutionary computation method that is inspired on the emulation of the collaborative behavior of social-spiders. In SSO, solutions imitate a set of spiders, which cooperate to each other based on the natural laws of the cooperative colony. Unlike the most popular evolutionary algorithms such as GA [16], PSO [17], HS [18], GSA [19] and CS [20], it explicitly evades the concentration of individuals in the best positions, avoiding critical flaws such as the premature convergence to sub-optimal solutions and the limited balance of exploration–exploitation. Such characteristics have motivated the use of SSO to solve an extensive variety of engineering applications such as energy theft detection [21], machine learning [22], electromagnetics [23], image processing [24] and integer programming problems [25].

This chapter presents a method for the optimal parameter calibration of fractional FCs based on the SSO algorithm. Under this approach, the calibration process is transformed into a multidimensional optimization problem where fractional orders, as well as controller parameters of the fuzzy system, are considered as a candidate solution to the calibration task. In the method, the SSO algorithm searches the entire parameter space while an objective function evaluates the performance of each parameter set. Conducted by the values of such objective function, the group of candidate solutions are evolved through the SSO algorithm so that the optimal solution can be found. Experimental evidence shows the effectiveness and robustness of the method for calibrating fractional FCs. A comparison with similar methods such as the Genetic Algorithms (GA), the Particle Swarm Optimization (PSO), the Harmony Search (HS), the Gravitational Search Algorithm (GSA) and the Cuckoo Search (CS) on different dynamical systems has been incorporated to demonstrate the performance of this approach. Conclusions of the experimental comparison are validated through statistical tests that properly support the discussion.

The Chapter is organized as follows: Sect. 3.2 introduces the concepts of fractional order systems; Sect. 3.3 describes the fractional fuzzy controller used in the calibration; Sect. 3.4 presents the characteristics of SSO; Sect. 3.5 formulates the parameter calibration problem; Sect. 3.6 shows the experimental results while some final conclusions are discussed in Sect. 3.7.

3.2 Fractional-Order Models

Dynamical fractional-order systems are modeled by using differential equations, which involve non-integer integral and/or derivative operators [26, 27]. Since these operators produce irrational continuous transfer functions, or infinite dimensional discrete transfer functions, fractional models are normally studied through simulation tools instead of analytical methods [5, 28–33]. The remainder of this section provides a background of the fundamental aspects of the fractional calculus, and the discrete integer-order approximations of fractional order operators that are used in this paper.

3.2.1 Fractional Calculus

Fractional calculus is a generalization of integration and differentiation to the non-integer order fundamental operator. The differential-integral operator, denoted by ${}_aD_t^\alpha$, takes both the fractional derivative and the fractional integral into a single expression defined as:

$$
{}_aD_t^\alpha = \begin{cases} \frac{d^\alpha}{dt^\alpha}, & \alpha > 0, \\ 1, & \alpha = 0, \\ \int_a^t (d\tau)^\alpha, & \alpha < 0. \end{cases} \tag{3.1}
$$

where a and t represent the operation bounds, whereas $\alpha \in \Re$. The commonly used definitions for fractional derivatives are the Grünwald-Letnikov, Riemann-Liouville [34] and Caputo [35]. According to the Grünwald-Letnikov approximation, the fractional-order derivative of order α is defined as follows:

$$
D_t^\alpha f(t) = \lim_{h \to 0} \frac{1}{h^\alpha} \sum_{j=0}^{\infty} (-1)^j \binom{\alpha}{j} f(t - jh) \tag{3.2}
$$

In the numerical calculation of fractional-order derivatives, the explicit numerical approximation of the α-th derivative at the points $kh, (k = 1, 2, \ldots)$ maintains the following form [36]:

$$
{}_{(k - M_m/h)}D_{t_k}^\alpha f(t) \approx h^{-\alpha} \sum_{j=0}^{k} (-1)^j \binom{\alpha}{j} f(t_k - j) \tag{3.3}
$$

where M_m is the memory length $t_k = kh$, h is the time step and $(-1)^j \binom{\alpha}{j}$ are the binomial coefficients. For their calculation, we can use the following expression:

$$
c_0^{(\alpha)} = 1, \quad c_j^{(\alpha)} = \left(1 - \tfrac{1+\alpha}{j}\right) c_{j-1}^{(\alpha)} \tag{3.4}
$$

Then, the general numerical solution of the fractional differential equation is defined as follows:

$$
y(t_k) = f(y(t_k), t_k) h^\alpha - \sum_{j=1}^{k} c_j^{(\alpha)} y(t_{k-j}) \tag{3.5}
$$

3.2.2 Approximation of Fractional Operators

Assuming zero initial conditions, the fractional operator is defined in the Laplace domain as $L({}_aD_t^\alpha f(t)) = s^\alpha F(s)$. Several approaches [37–39] have been proposed for producing discrete versions of continuous operators of type s^α. In this chapter, the Grünwald-Letnikov approximation has been used due to its interesting properties for generating discrete equivalences [40–43]. Under this method, the discretization considers the following model:

$$D^\alpha(z^{-1}) = \left(\frac{1-z^{-1}}{T_c}\right) = \sum_{k=0}^{\infty}\left(\frac{1}{T_c}\right)^\alpha(-1)\binom{\alpha}{k}z^{-k} = \sum_{k=0}^{\infty}h^\alpha(k)z^{-k}, \quad (3.6)$$

where $h^\alpha(k)$ is the impulse response sequence, whereas T_c represents the sampling frequency. It has been already demonstrated in the literature [36] that rational models converge faster than polynomial methods. Consequently, the Padé approximation approach has been employed to obtain a fractional model from the impulse response by using the definition provided in Eq. 3.7.

$$H(z^{-1}) = \frac{b_0 + b_1z^{-1} + \cdots + b_mz^{-m}}{1 + a_1z^{-1} + \cdots + a_nz^{-n}} = \sum_{k=0}^{\infty}h(k)z^{-k}, \quad m \leq n \quad (3.7)$$

where m, n and the parameters a_i and b_i are calculated by adjusting the first $m + n + 1$ coefficients of $h^\alpha(k)$.

3.3 Fuzzy Controller

A fuzzy controller (FC) is a nonlinear system produced from empirical rules. Such empirical information may come from a human operator who directly manipulates the process. Each rule, just as the natural language, presents an IF-THEN format. The collection of all rules constitutes the rule base that emulates the decision-making process of the operator. An important characteristic of one FC is the partitioning of the control scheme into regions [44]. At each region, the control strategy can be simply modeled by using a rule that associates the region under which certain actions are performed. Despite proposing several configurations of FCs in the literature, the fuzzy fractional $\text{PD}^\alpha + \text{I}$ has been selected since it presents interesting characteristics of robustness and stability [5]. In this structure, the integral error is incorporated to the output of the fuzzy fractional PD^α controller. Under this configuration, the integral action supports the elimination of the final steady state error.

The controller configuration is shown in Fig. 3.1. In the Figure, *E*, *DE* and *IE* represents the error, the fractional derivative error and the integral error, respectively. It has four gains K_p, K_d, K_i and K_u to be calibrated, the first three gains

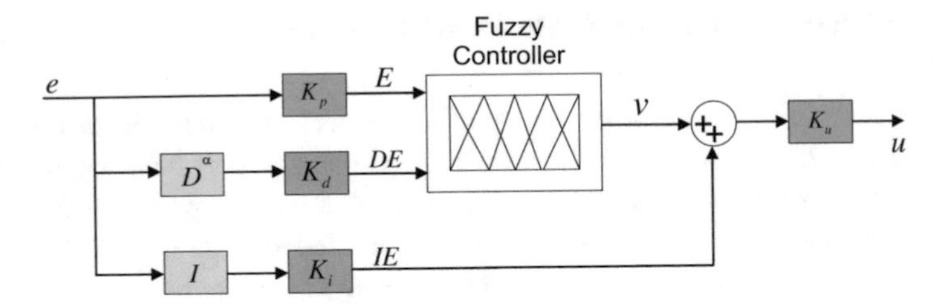

Fig. 3.1 Fuzzy PD$^\alpha$ +I controller

Table 3.1 Rule base of the controller to be calibrated

E/DE	NL	NM	NS	ZR	PS	PM	PL
NL	NL	NL	NL	NL	NM	NS	ZR
NM	NL	NL	NL	NM	NS	ZR	PS
NS	NL	NL	NM	NS	ZR	PS	PM
ZR	NL	NM	NS	ZR	PS	PM	PL
PS	NM	NS	ZR	PS	PM	PL	PL
PM	NS	ZR	PS	PM	PL	PL	PL
PL	ZR	PS	PM	PL	PL	PL	PL

correspond to the input and the last one to the output. The control action u is a nonlinear mapping function of E, DE and IE with the following model:

$$\begin{aligned} u(k) &= (f(E, CE) + IE)K_u \\ u(k) &= \left[f\left(K_p e, K_d D^\alpha e\right) + K_i Ie\right] \cdot K_u, \end{aligned} \tag{3.8}$$

A fuzzy controller consists of three conceptual components: a rule base, which contains a selection of fuzzy rules; a database, which defines the membership functions used by the fuzzy rules; and a reasoning mechanism, which performs the inference procedure. There are two different fuzzy systems: the Mamdani [45] and the Takagi-Sugeno (TS) [46]. In order to maintain compatibility to similar works reported in the literature, the rule base and the membership functions are selected the same as [5, 9]. Under such conditions, Table 3.1 shows the rule base used by the fuzzy controller to be calibrated. In Table 3.1, **NL**, **NM**, **NS**, **ZR**, **PS**, **PM** and **PL** represent the linguistic variables "Negative Large", "Negative Medium", "Negative Small", "Zero", "Positive Small", "Positive Medium" and "Positive Large", respectively. Figure 3.2 shows the membership functions that model the premises and the consequences of each rule. Consequently, a determined rule from Table 3.1 can be constructed in the following form:

If ***E*** *is* **NL** and ***DE*** is **ZR** then ***v*** is **NL**

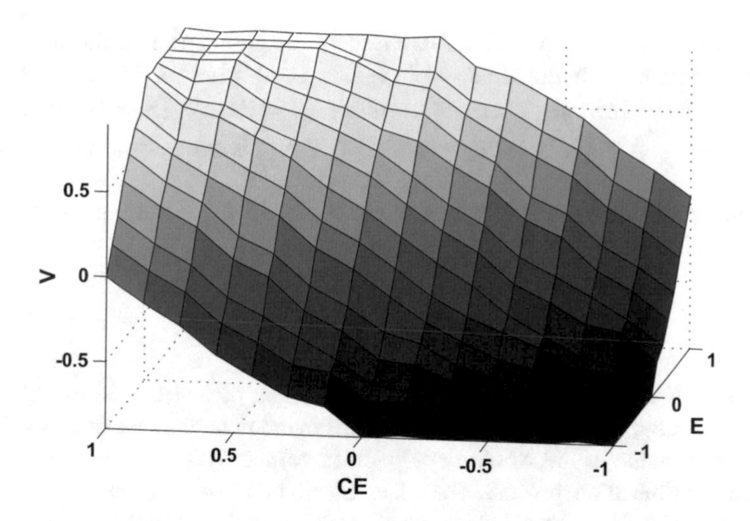

Fig. 3.2 Control surface

In this rule, the control strategy can be simply modeled as follows: if the error is "Negative Large" and the error derivate "Zero" then the output is "Negative Large". The acting of all rules produces the control strategy which is shown by the non-linear surface in Fig. 3.2.

3.4 Social Spider Optimization (SSO)

The social spider optimization (SSO) algorithm [15] is an evolutionary computation method that emulates the cooperative behavior of spiders within a communal colony. SSO has been designed to find the global solution of a nonlinear optimization problem with box constraints in the form:

$$\begin{array}{ll} \text{minimize} & f(\mathbf{x}) \quad \mathbf{x} = (x^1, \ldots, x^d) \in \mathbb{R}^d \\ \text{subject to} & \mathbf{x} \in \mathbf{X} \end{array} \tag{3.9}$$

where $f : \mathbb{R}^d \rightarrow \mathbb{R}$ is a nonlinear function whereas $\mathbf{X} = \left\{\mathbf{x} \in \mathbb{R}^d \middle| l_h \leq x^h \leq u_h, h = 1, \ldots, d\right\}$ is a bounded feasible space, constrained by the lower (l_h) and upper (u_h) limits.

SSO utilizes a population $\mathbf{S}$ of N candidate solutions to solve the problem formulated in Eq. 3.1. Each candidate solution represents a spider position whereas the general web symbolizes the search space $\mathbf{X}$. In SSO, the spider population $\mathbf{S}$ is classified into two categories: males $(\mathbf{M})$ and females $(\mathbf{F})$. In order to simulate a real spider colony, in SSO, the number N_f of females $\mathbf{F}$ is randomly selected within a range of 65–90% of the entire population $\mathbf{S}$, whereas the rest N_m is considered as male individuals ($N_m = N - N_f$). Under such conditions, the Group **F** assembles the set of female individuals ($\mathbf{F} = \{\mathbf{f}_1, \mathbf{f}_2, \ldots, \mathbf{f}_{N_f}\}$) whereas **M** groups the male members ($\mathbf{M} = \{\mathbf{m}_1, \mathbf{m}_2, \ldots, \mathbf{m}_{N_m}\}$), where $\mathbf{S} = \mathbf{F} \cup \mathbf{M}$ ($\mathbf{S} = \{\mathbf{s}_1, \mathbf{s}_2, \ldots, \mathbf{s}_N\}$), such that $\mathbf{S} = \{\mathbf{s}_1 = \mathbf{f}_1, \mathbf{s}_2 = \mathbf{f}_2, \ldots, \mathbf{s}_{N_f} = \mathbf{f}_{N_f}, \mathbf{s}_{N_f+1} = \mathbf{m}_1, \mathbf{s}_{N_f+2} = \mathbf{m}_2, \ldots, \mathbf{s}_N = \mathbf{m}_{N_m}\}$.

In the approach, each spider i maintain a weight w_i according to its solution quality. Therefore, w_i is calculated as follows:

$$w_i = \frac{fitness_i - worst}{best - worst} \tag{3.10}$$

where $fitness_i$ represents the fitness value produced by the evaluation of the i-th spider's position, $i \in 1, \ldots, N$. *best* and *worst* symbolize the best fitness value and worst fitness value of the whole population **S**, respectively.

In the optimization process, the main mechanism of SSO is the information exchange, which it is simulated trough vibrations produced in the communal web. The vibration that a spider i perceives from a spider j is modeled with the following expression:

$$V_{i,j} = w_j e^{d_{i,j}^2} \tag{3.11}$$

where w_j represents the weight of the spider j and $d_{i,j}^2$ the distance between both spiders. It is considered that each spider i is only able to perceive three types of vibrations $V_{i,c}$, $V_{i,b}$ and $V_{i,f}$.

$V_{i,c}$ is the vibration transmitted by the nearest individual c with a higher weight with regard to $i(w_c > w_i)$. $V_{i,b}$ represents the vibration emitted by best element of the entire population **S**. Finally, $V_{i,f}$ considers the vibration produced by the nearest female spider. This vibration type is only applicable if i is a male individual.

In the operation of SSO, a population of N spiders is processed from the initial stage ($k = 0$) to a determined number *gen* of iterations ($k = gen$). Each individual depending on its gender is conducted by a set of different evolutionary operators. Therefore, in case of the female members, a new position $\mathbf{f}_i^{k+1}$ is generated by modifying the current element location $\mathbf{f}_i^k$. The modification is randomly controlled by using a probability factor *PF*. Consequently, the movement is produced in relation to other spiders according their vibrations, which are transmitted trough the communal web:

$$\mathbf{f}_i^{k+1} = \begin{cases} \mathbf{f}_i^k + \alpha \cdot V_{i,c} \cdot (\mathbf{s}_c - \mathbf{f}_i^k) + \beta \cdot V_{i,b} \cdot (\mathbf{s}_b - \mathbf{f}_i^k) + \delta \cdot (\text{rand} - \frac{1}{2}) & \text{with probability } PF \\ \mathbf{f}_i^k - \alpha \cdot V_{i,c} \cdot (\mathbf{s}_c - \mathbf{f}_i^k) - \beta \cdot V_{i,b} \cdot (\mathbf{s}_b - \mathbf{f}_i^k) + \delta \cdot (\text{rand} - \frac{1}{2}) & \text{with probability} 1 - PF \end{cases} \tag{3.12}$$

here α, β, δ and *rand* represent random numbers between [0,1] whereas k is the iteration number. The individuals $\mathbf{s}_c$ and $\mathbf{s}_b$ symbolize the nearest member to i that maintains a higher weight and the best element of the complete population $\mathbf{S}$, respectively.

On the other hand, male spider members are classified into two types: non-dominant (**ND**) and dominant (**D**). The dominant group **D** is composed by the half of the male individuals whose fitness values are better with regard to the complete male set. Consequently, the non-dominant (**ND**) category collects the rest of the male elements. In the optimization process, male members are operated according to the following model:

$$\mathbf{m}_i^{k+1} = \begin{cases} \mathbf{m}_i^k + \alpha \cdot V_{i,f} \cdot (\mathbf{s}_f - \mathbf{m}_i^k) + \delta \cdot (\text{rand} - \frac{1}{2}) & \text{if } \mathbf{m}_i^k \in \mathbf{D} \\ \mathbf{m}_i^k + \alpha \cdot \left(\frac{\sum_{h \in \mathbf{ND}} \mathbf{m}_h^k \cdot w_h}{\sum_{h \in \mathbf{ND}} w_h} - \mathbf{m}_i^k \right) & \text{if } \mathbf{m}_i^k \in \mathbf{ND} \end{cases} \tag{3.13}$$

where $\mathbf{s}_f$ symbolizes the nearest female element to the male individual i.

The final operation in SSO is mating. It is performed between dominant males and the female individuals. Under this operation, a new individual $\mathbf{s}_{new}$ is produced by the combination of a dominant male $\mathbf{m}_g$ and other female members within a specific range r. The weight of each involved element defines the probability of influence of each spider into $\mathbf{s}_{new}$. The elements with heavier weights are more likely to influence the new individual $\mathbf{s}_{new}$. Once $\mathbf{s}_{new}$ is generated, it is compared with the worst element of the colony. If $\mathbf{s}_{new}$ is better than the worst spider, the worst spider is replaced by $\mathbf{s}_{new}$. Otherwise, $\mathbf{s}_{new}$ is discarded. Figure 3.3 illustrates the operations of the optimization process performed by the SSO algorithm. More details can be found in [15].

3.5 Problem Formulation

In the design stage of fractional FCs, the parameter calibration process is transformed into a multidimensional optimization problem where fractional orders, as well as controller parameters of the fuzzy system, are both considered as decision variables. Under this approach, the complexity of the optimization problem tends to produce multimodal error surfaces whose cost functions are significantly difficult to minimize.

This chapter presents an algorithm for the optimal parameter calibration of fractional FCs. To determine the parameters, the estimation method uses the Social Spider Optimization (SSO) method. Different to the most of existent evolutionary algorithms, the method explicitly evades the concentration of individuals in the best positions, avoiding critical flaws such as the premature convergence to sub-optimal solutions and the limited balance of exploration–exploitation.

Therefore, the calibration process consists of finding the optimal controller parameters that present the best possible performance for the regulation of a dynamical system. Figure 3.4 illustrates the SSO scheme for the parameter calibration process.

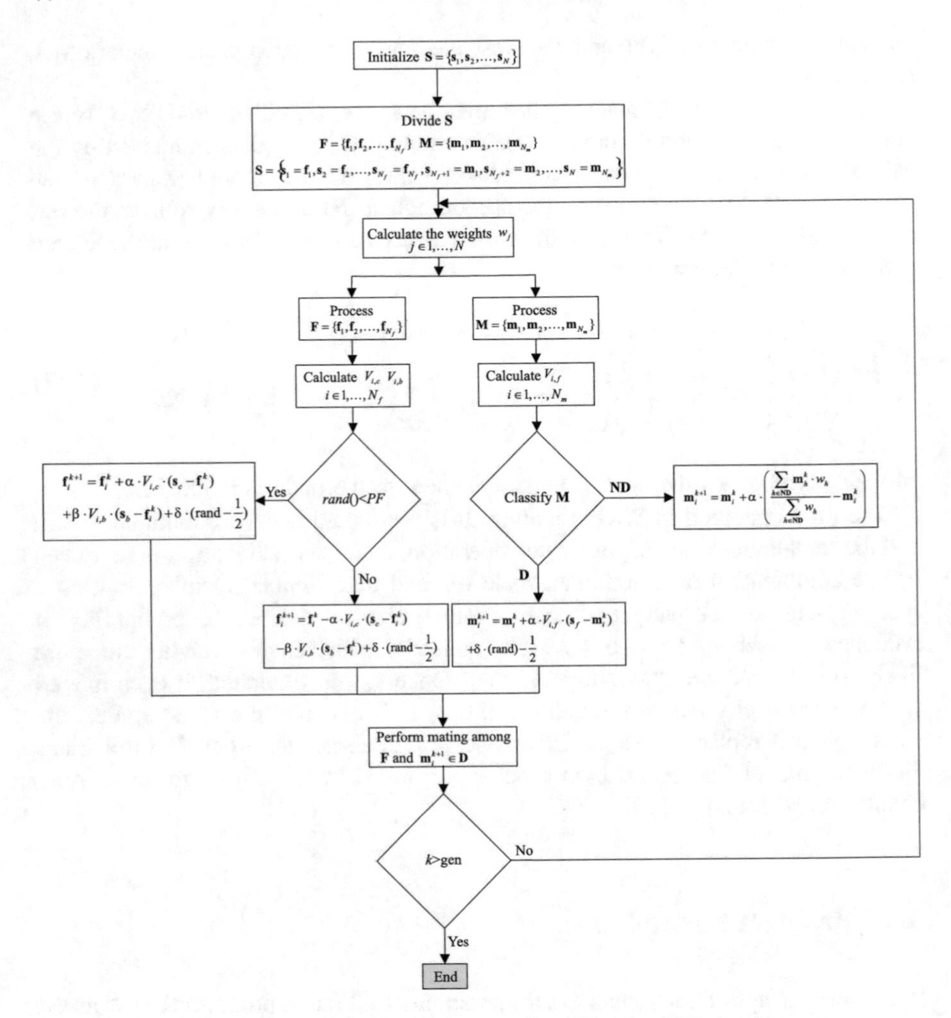

Fig. 3.3 Operations of the optimization process performed by the SSO algorithm

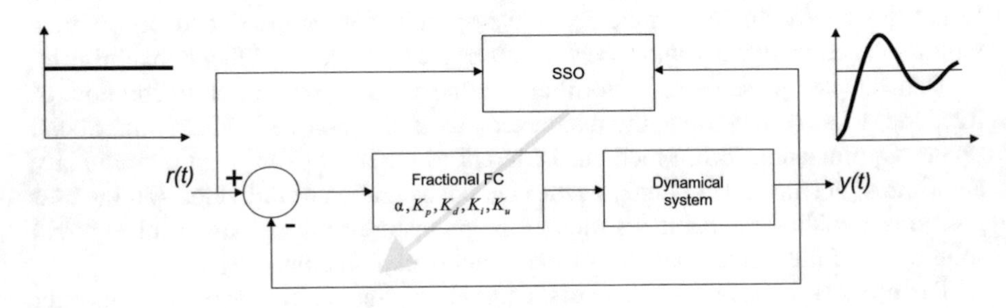

Fig. 3.4 SSO scheme for the parameter calibration process

Under such conditions, the fractional fuzzy controller parameters $(\alpha, K_p, K_d, K_i, K_u)$ represent the dimensions of each candidate solution (spider position) for the calibration problem. To evaluate the performance of the fractional fuzzy controller under each parameter configuration (candidate solution), the Integral Time Absolute Error (ITAE) [47] criterion has been considered. The ITAE index J measures the similarity between the closed-loop step response $y(t)$ produced by a determined parameter configuration $(\alpha, K_p, K_d, K_i, K_u)$ and the step function $r(t)$. Therefore, the quality of each candidate solution is evaluated according to the following model:

$$J(\alpha, K_p, K_d, K_i, K_u) = \int_0^\infty t|r(t) - y(t)| \tag{3.14}$$

Thereby, the problem of parameter calibration can be defined by the following optimization formulation:

$$\begin{array}{lll} \text{minimize} & J(\mathbf{x}) & \mathbf{x} = (\alpha, K_p, K_d, K_i, K_u) \in \mathbb{R}^5 \\ \text{subject to} & & 0 \leq \alpha \leq 3 \\ & & 0 \leq K_p \leq 5 \\ & & 0 \leq K_d \leq 5 \\ & & 0 \leq K_i \leq 5 \\ & & 0 \leq K_u \leq 5 \end{array} \tag{3.15}$$

3.6 Numerical Simulations

This section presents the performance of the SSO scheme for the calibration of fractional FCs considering several dynamical systems. The algorithm is also evaluated in comparison to other similar approaches that are based on evolutionary algorithms. To test the performance of the SSO approach, the technique uses a representative set of three transfer functions that have been previously employed. Equation 3.4–3.6 present the transfer functions that are used in our simulations. Such functions involve three different system categories: High-order plants $(G_1(s))$, non-minimum systems $(G_2(s))$ and dynamical fractional systems $(G_3(s))$.

$$G_1(s) = \frac{1}{(s+1)(1+0.5s)(1+0.25s)(1+0.125s)} \tag{3.16}$$

$$G_2(s) = \frac{1-5s}{(s+1)^3} \tag{3.17}$$

$$G_3(s) = \frac{1}{(s^{1.5} + 1)} \tag{3.18}$$

In the experiments, we have applied the SSO algorithm to calibrate the fractional parameters for each dynamical systems, and the results are compared to those produced by the Genetic Algorithms (GA) [5], Particle Swarm Optimization (PSO) [9], Harmony Search (HS) [10], Gravitational Search Algorithm (GSA) [11] and Cuckoo Search (CS) [12]. In the comparison, all methods have been set according to their own reported guidelines. Such configurations are described as follows:

1. PSO, parameters $c_1 = 2$, $c_2 = 2$ and weights factors have been set to $w_{\max} = 0.9$, and $w_{\min} = 0.4$ [17].
2. GA, the crossover probability is 0.55, the mutation probability is 0.10 and number of elite individuals is 2. Furthermore, the roulette wheel selection and the 1-point crossover are both applied.
3. GSA, From the model $G_t = G_O e^{-\alpha \frac{t}{T}}$, it is considered $\alpha = 10$, $G_O = 100$ and $T = 100$ or $T = 500$.
4. HS, its parameters are set as follows: the harmony memory consideration rate $HMCR = 0.7$, the pitch adjustment rate $PAR = 0.3$ and the Bandwidth rate $BW = 0.1$.
5. CS, its elements are configured such as the discovery rate $p_a = 0.25$ and the stability index $\beta = 3/2$.
6. SSO, the parameter PF has been set to 0.7 following an experimental definition.

The experimental results are divided into three sub-sections. In the first Sect. (3.6.1), the performance of the SSO algorithm is evaluated with regard to high-order plants $(G_1(s))$. In the second Sect. (3.6.2), the results for non-minimum systems $(G_2(s))$ are provided and finally, in the third Sect. (3.6.3), the performance of the calibration scheme over fractional dynamical systems $(G_3(s))$ is discussed.

3.6.1 Results Over High-Order Plants $(G_1(s))$

In this experiment, the performance of the SSO calibration scheme is compared to GA, PSO, HS, GSA and CS, considering the regulation of high-order dynamical systems $(G_1(s))$. In the simulations, a temporal response from 0 to 10 s has been considered. In the comparison, all algorithms are operated with a population of 50 individuals (**N** = 50). To appropriately evaluate the convergence properties of all calibration methods, the maximum number of generations has been set to (A) 100 iterations and (B) 500 iterations. This stop criterion has been selected to maintain compatibility to similar works reported in the literature [5, 9, 28]. By selecting such number of iterations, the experiment aims to test the quality of the produced

Table 3.2 Calibrated parameters for $G_1(s)$ produced by each algorithm after 100 iterations

$G_1(s)$	K_p	K_d	K_i	K_u	α	ITAE
PSO	0.3034	0	0.4475	2.9988	0	5281.2115
GA	0.7581	0.3510	0.3038	4.3276	0.8000	926.1352
GSA	1.3387	2.7209	0.5482	1.0545	0.2311	4164.1935
HS	0.7867	0.8128	0.8271	3.6129	0.9319	3562.1834
CS	0.9700	0.3497	0.4054	3.0002	0.9516	916.5816
SSO	0.8100	0.3493	0.2392	4.8235	0.9897	492.2912

Table 3.3 Calibrated parameters for $G_1(s)$ produced by each algorithm after 500 iterations

$G_1(s)$	K_p	K_d	K_i	K_u	α	ITAE
PSO	0	0.5061	0.4681	4.0578	0.4629	2900.7502
GA	1.1860	0.3826	0.5204	1.5850	0.9497	974.0881
GSA	1.2000	0.6531	0.9607	2.5442	0.8745	1975.3254
HS	1.3112	0.9450	0.9262	1.2702	0.6075	2776.2160
CS	1.0093	0.4506	0.2611	4.6964	1.0002	464.5376
SSO	1.0386	0.4621	0.2751	4.4147	0.9998	473.7492

solutions when the operation of each calibration method is limited to a reduced number of iterations.

All the experimental results in this section consider the analysis of 35 independent executions of each algorithm. Table 3.2 presents the calibrated parameters obtained through each method. Such results consider the best controller parameters in terms of the produced ITAE values after 100 iterations. On the other hand, Table 3.3 shows the calibrated parameters considering 500 iterations.

According to Tables 3.2 and 3.3, the SSO scheme provides better performance than GA, PSO, HS, GSA and CS for both cases. Such differences are directly related to a better trade-off between exploration and exploitation of the SSO method. It is also evident that the SSO method produces similar results with 100 or 500 iterations. Therefore, it can be established that SSO maintains better convergence properties than GA, PSO, HS, GSA and CS in the process of parameter calibration.

Figure 3.5 exhibits the step responses produced by each parameter set, considering 100 and 500 iterations. The remarkable convergence rate of the SSO algorithm can be observed at Fig. 3.5. According to the graphs, the step responses produced by the controller parameters that have been defined through SSO, are practically the same irrespective of the number of iterations. This fact means that the SSO scheme is able to find an acceptable solution in less than 100 iterations.

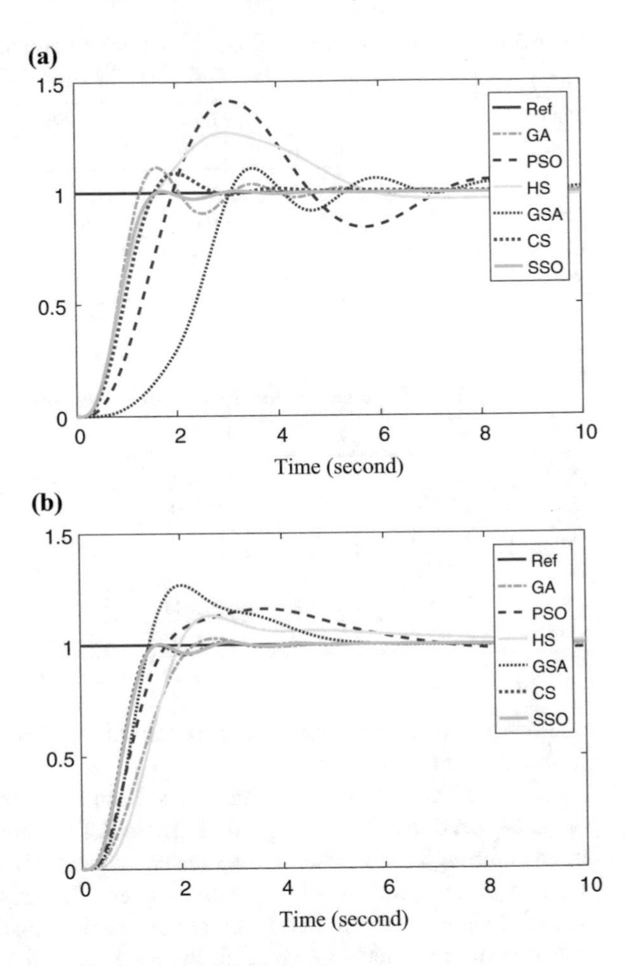

Fig. 3.5 Step responses after applying the calibrated parameters to the high order plant $G_1(s)$ with **a** 100 iterations, and with **b** 500 iterations

3.6.2 Results Over Non-minimum Systems $(G_2(s))$

This section presents the comparison of the SSO calibration scheme with GA, PSO, HS, GSA and CS, considering the regulation of non-minimum systems $(G_2(s))$. Non-minimum systems are defined by transfer functions with one or more poles or zeros in the right half of the s-plane. As a consequence, the response of a non-minimum system to a step input exhibits an "undershoot", which indicates that the output of the dynamical system becomes negative first before changing direction to positive values.

The experimental simulation runs from 0 to 50 s. All algorithms are operated with a population of 50 individuals (**N** = 50), matching with the experiments in Sect. 3.6.1. Table 3.4 presents the calibrated parameter of each method after 100 iterations, while Table 3.5 exhibits the results for 500 iterations. Both tables show that the SSO scheme delivers better results than GA, PSO, HS, GSA and CS in

Table 3.4 Calibrated parameters for $G_2(s)$ produced by each algorithm after 100 iterations

$G_2(s)$	K_p	K_d	K_i	K_u	α	ITAE
PSO	0.4645	0	0.2378	0.4147	0.0643	50289.0994
GA	0.6061	0.0326	0.3175	0.2909	0.2000	55043.6316
GSA	0.9377	1.8339	0.5020	0.1427	1.0266	101160.6241
HS	2.2838	3.7685	0.1328	0.5324	2.3214	126996.7047
CS	0.9305	1.1329	1.1045	0.0674	0.0222	90962.6199
SSO	0.4668	0.1165	0.2139	0.4642	0.5470	44368.6620

Table 3.5 Calibrated parameters for $G_2(s)$ produced by each algorithm after 500 iterations

$G_2(s)$	K_p	K_d	K_i	K_u	α	ITAE
PSO	0.4606	0	0.2027	0.4866	0.0688	46912.4985
GA	1.0449	1.1921	0.8839	0.0903	0.0822	81550.0790
GSA	0.8862	1.3919	0.3746	0.2386	1.9259	63186.5783
HS	1.0362	1.1105	0.5360	0.1389	0.9007	91536.3894
CS	0.0386	0.0059	0.0243	4.0625	0.5516	43565.1588
SSO	0.4537	0.1597	0.2004	0.5124	0.6422	41772.3344

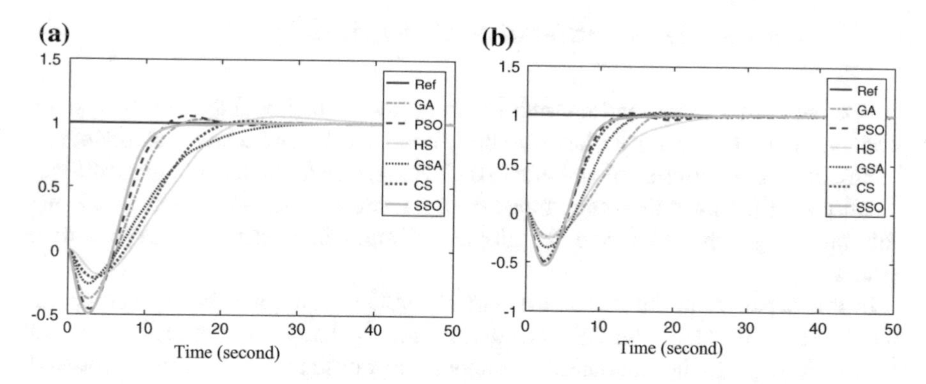

Fig. 3.6 Step responses after applying the calibrated parameters to the high order plant $G_2(s)$ with **a** 100 iterations, and with **b** 500 iterations

terms of the ITAE index. Figure 3.6 presents the step responses produced by each parameter set, considering 100 and 500 iterations. By analyzing the plot in Fig. 3.6, it is observed that the step response of the SSO scheme is less sensitive to the number of iterations than other techniques.

Table 3.6 Calibrated parameters for $G_3(s)$ produced by each algorithm after 100 iterations

$G_3(s)$	K_p	K_d	K_i	K_u	α	ITAE
PSO	1.331	0	0.6937	5	5	311.4558
GA	1.3329	0.6341	0.6130	5	0.4932	97.7016
GSA	1.0823	0.6463	0.2924	4.0152	0.5802	346.6765
HS	0.7867	0.8128	0.8271	3.6129	0.9319	3562.1834
CS	1.2220	0.6590	0.6647	5	0.4232	105.0266
SSO	1.3173	0.6560	0.5932	4.9797	0.5091	98.5974

Table 3.7 Calibrated parameters for $G_3(s)$ produced by each algorithm after 500 iterations

$G_3(s)$	K_p	K_d	K_i	K_u	α	ITAE
PSO	1.0187	1.2010	0.7553	5	0	181.5380
GA	1.3320	0.5599	0.5991	5	0.5631	97.1981
GSA	1.3319	0.7502	0.6689	3.4968	0.5185	152.2198
HS	1.3112	0.9450	0.9262	1.2702	0.6075	2776.2160
CS	1.3325	0.5857	0.5987	4.9999	0.5450	97.0307
SSO	1.3087	0.5883	0.5808	4.9991	0.5642	97.1085

3.6.3 *Results Over Fractional Systems* $(G_3(s))$

Unlike high-order plants and non-minimum systems, fractional dynamical systems produce multimodal error surfaces with different local optima. As a consequence, fractional fuzzy controllers that regulate their behavior are, in general, more difficult to calibrate [9]. Under such conditions, the experiment reflects the capacity of each calibration algorithm to locate the global optimum in presence of several local optima.

In this experiment, the performance of the SSO calibration scheme is compared to GA, PSO, HS, GSA and CS, considering the regulation of fractional dynamical systems $(G_3(s))$. In the simulations, a temporal response from 0 to 3 s is considered. In the test, all algorithms are operated with a population of 50 individuals (**N** = 50).

The calibrated parameters are averaged over 30 executions obtaining the values reported in Tables 3.6 and 3.7. The results exhibit the configuration for each method with 100 and 500 iterations, respectively. It is evident that the SSO scheme presents better performance than PSO, HS and GSA independently on the number of iterations. However, the difference between GA and the SSO approach in terms of the ITAE index is relatively small for the case of 100 iterations. On the other hand, in the case of 500 iterations, the performance among the SSO approach, GA and CS are practically the same. Figure 3.7 presents the step response that is produced by each parameter set, considering 100 and 500 iterations. Similar to Sects. 6.1 and 6.2, it is demonstrated (from Fig. 3.7) that the SSO-calibrator obtains better solutions than GA, HS and PSO yet demanding a lower number of iterations.

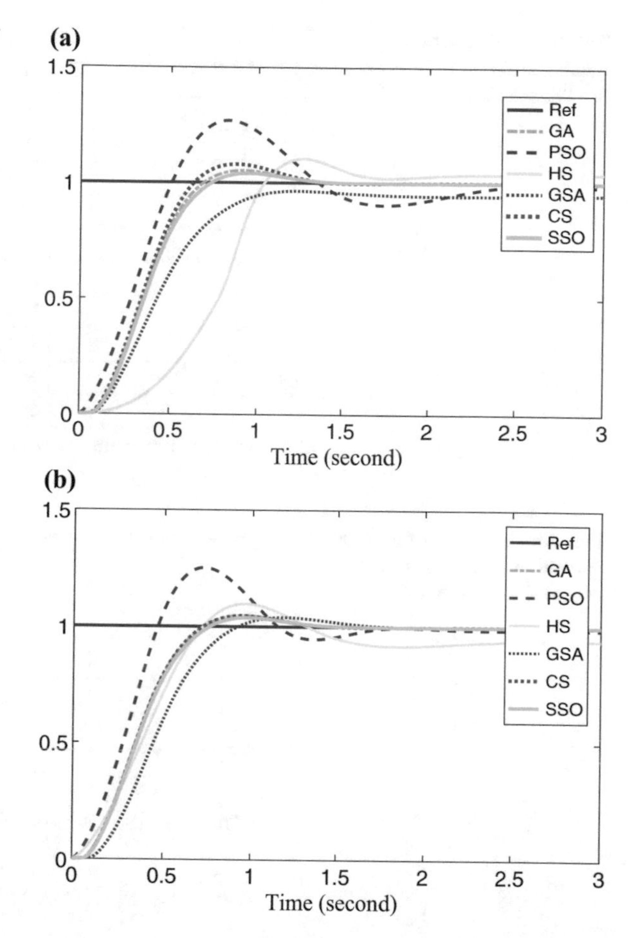

Fig. 3.7 Step responses after applying the calibrated parameters to the high order plant $G_3(s)$ with **a** 100 iterations, and with **b** 500 iterations

Finally, in order to stress the importance of the fractional $\mathrm{PD}^{\alpha}+\mathrm{I}$ scheme, an experiment that evaluates the influence of the parameter α in the regulation of $G_3(s)$ is conducted. In the experiment, the fractional $\mathrm{PD}^{\alpha}+\mathrm{I}$ controller is operated as an integer fuzzy controller by setting $\alpha = 1$. Under such conditions, the rest of the parameters of $\mathrm{PD}^{\alpha}+\mathrm{I}$ (K_p, K_d, K_i, K_u) are calibrated through the SSO approach considering the regulation of the fractional system $G_3(s)$. Then, the values α vary from 0 to 1 while registering the performance of the regulation.

As a result of the optimization method, the following parameter values are obtained: $(K_p, K_d, K_i, K_u) \equiv (1.3087, 0.5883, 0.5808, 4.0012)$ with ITAE = 418.8032. After calibrating the integer fuzzy controller, the values of α are modified from 0 to 1, while the parameter set remains fixed to $(1.3087, 0.5883, 0.5808, 4.0012)$. Table 3.8 presents the results obtained from the experiment. Such values report the regulation quality of $G_3(s)$ in terms of the ITEA values. By analyzing Table 3.8, it is clear that the regulation quality strongly depends on the selection of the order for α. Particularly in

Table 3.8 Regulation quality of $G_3(s)$ in terms of the ITEA values

α	ITAE
0	216.839600
0.1	198.666000
0.2	180.783670
0.3	159.871905
0.4	134.127486
0.5	108.724153
0.6	**105.942730**
0.7	136.404140
0.8	183.213124
0.9	262.521840
1	418.803215

Bold elements represent the best values

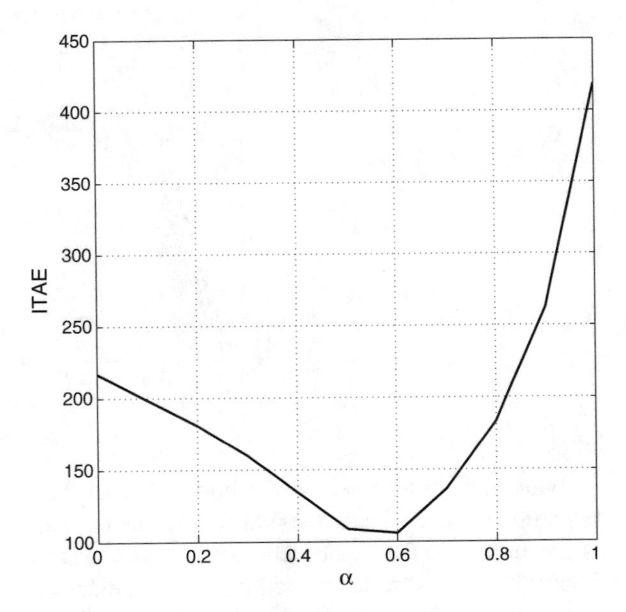

Fig. 3.8 Influence of α in the regulation quality of $G_3(s)$ in terms of the ITEA values

this experiment, the best regulation performance is reached when the order of α is set to 0.6. Figure 3.8 presents the influence of α in terms of the regulation quality.

3.7 Conclusions

Due to its multiple applications, the calibration of fractional fuzzy controllers has attracted the interests of several research communities. In the calibration process, the parameter estimation is transformed into a multidimensional optimization problem whose fractional order and the corresponding controller parameters of the fuzzy system are considered as decision variables. Under this approach, the complexity of fractional-order chaotic systems tends to produce multimodal error surfaces for which their cost functions are significantly difficult to minimize. Several algorithms that are based on evolutionary computation principles have been successfully applied to calibrate the parameters of fuzzy systems. However, most of them still have an important limitation since they frequently obtain sub-optimal results as a consequence of an inappropriate balance between exploration and exploitation in their search strategies.

This chapter presents a method for the optimal parameter calibration of fractional FCs that is based on the SSO algorithm. The SSO algorithm is a novel evolutionary computation method that is inspired on the emulation of the collaborative behavior of social-spiders. Unlike most of the existing evolutionary algorithms, the method explicitly evades the concentration of individuals in best positions, avoiding critical flaws such as the premature convergence to sub-optimal solutions and the limited balance of exploration–exploitation.

In order to illustrate the proficiency and the robustness of this approach, SSO scheme has been experimentally evaluated considering three different system categories: high-order plants, non-minimum systems and dynamical fractional systems. To assess the performance of the SSO algorithm, it has been compared to other similar evolutionary approaches such as Genetic Algorithms (GA), Particle Swarm Optimization (PSO), Harmony Search (HS), Gravitational Search Algorithm (GSA) and Cuckoo Search (CS). The experiments have demonstrated that the SSO method outperforms other techniques in terms of solution quality and convergence.

References

1. Oldham, K.B., Spanier, J.: The Fractional Calculus: Theory and Application of Differentiation and Integration to Arbitrary Order. Academic Press, New York (1974)
2. Podlubny, I.: Fractional Differential Equations. Academic Press, San Diego (1999)
3. Das, S., Pan, I., Das, S., Gupta, A.: A novel fractional order fuzzy PID controller and its optimal time domain tuning based on integral performance indices. J. Eng. Appl. Artif. Intell. **25**, 430–442 (2012)
4. Delavari, H., Ghaderi, R., Ranjbar, A., Momani, S.: Fuzzy fractional order sliding mode controller for nonlinear systems. Commun. Nonlinear Sci. Numer. Simul. **15**(4), 963–978 (2010)
5. Jesus, I.S., Barbosa, R.S.: Genetic optimization of fuzzy fractional PD + I controllers. ISA Trans. **57**, 220–230 (2015)
6. Barbosa, R.S., Jesus, I.S.: A methodology for the design of fuzzy fractional PID controllers. In: ICINCO 2013—Proceedings of the 10th International Conference on Informatics in Control, Automation and Robotics, vol. 1, pp. 276–281 (2013)
7. Zadeh, L.A.: Fuzzy sets. Inf. Control **8**, 338–353 (1965)
8. He, Y., Chen, H., He, Z., Zhou, L.: Multi-attribute decision making based on neutral averaging operators for intuitionistic fuzzy information. Appl. Soft Comput. **27**, 64–76 (2015)
9. Pana, I., Das, S.: Fractional order fuzzy control of hybrid power system with renewable generation using chaotic PSO. ISA Trans. **62**, 19–29 (2016)

10. Roy, G.G., Chakraborty, P., Das, S.: Designing fractional-order PIλDμ controller using differential harmony search algorithm. Int. J. Bio-Inspired Comput. **2**(5), 303–309 (2010)
11. Xu, Y., Zhou, J., Xue, X., Fu, W., Zhu, W., Li, C.: An adaptively fast fuzzy fractional order PID control for pumped storage hydro unit using improved gravitational search algorithm. Energy Convers. Manage. **111**, 67–78 (2016)
12. Sharma, R., Rana, K.P.S., Kumar, V.: Performance analysis of fractional order fuzzy PID controllers applied to a robotic manipulator. Expert Syst. Appl. **41**, 4274–4289 (2014)
13. Tan, K.C., Chiam, S.C., Mamun, A.A., Goh, C.K.: Balancing exploration and exploitation with adaptive variation for evolutionary multi-objective optimization. Eur. J. Oper. Res. **197**, 701–713 (2009)
14. Chen, G., Low, C.P., Yang, Z.: Preserving and exploiting genetic diversity in evolutionary programming algorithms. IEEE Trans. Evol. Comput. **13**(3), 661–673 (2009)
15. Cuevas, E., Cienfuegos, M., Zaldívar, D., Pérez-Cisneros, M.: A swarm optimization algorithm inspired in the behavior of the social-spider. Expert Syst. Appl. **40**(16), 6374–6384 (2013)
16. Goldberg, D.E.: Genetic Algorithm in Search Optimization and Machine Learning. Addison-Wesley, Boston (1989)
17. Kennedy, J., Eberhart, R.: Particle swarm optimization. In: Proceedings of the 1995 IEEE International Conference on Neural Networks, vol. 4, pp. 1942–1948, December 1995
18. Geem, Z.W., Kim, J., Loganathan, G.: Music-inspired optimization algorithm harmony search. Simulation **76**, 60–68 (2001)
19. Rashedi, E., Nezamabadi-pour, H., Saryazdi, S.: GSA: a gravitational search algorithm. Inf. Sci. **179**, 2232–2248 (2009)
20. Yang, X.S., Deb, S.: Engineering optimization by cuckoo search. Int. J. Math. Model. Numer. Optim. **1**(4), 339–343 (2010)
21. Pereira, D.R., Pazoti, M.A., Pereira, L.A.M., Rodrigues, D., Ramos, C.O., Souza, A.N., Papa, J.P.: Social-spider optimization-based support vector machines applied for energy theft detection. Comput. Electr. Eng. **49**, 25–38 (2016)
22. Mirjalili, S.Z., Saremi, S., Mirjalili, S.M.: Designing evolutionary feedforward neural networks using social spider optimization algorithm. Neural Comput. Appl. (8), 1919–1928 (2015)
23. Klein, C.E., Segundo, E.H.V., Mariani, V.C., Coelho, L.D.S.: Modified social-spider optimization algorithm applied to electromagnetic optimization. IEEE Trans. Magn. **52**(3), 2–10 (2016)
24. Ouadfel, S., Taleb-Ahmed, A.: Social spiders optimization and flower pollination algorithm for multilevel image thresholding: a performance study. Expert Syst. Appl. **55**, 566–584 (2016)
25. Tawhid, M.A., Ali, A.F.: A simplex social spider algorithm for solving integer programming and minimax problems. Memetic Comput. (In press)
26. Jesus, I., Machado, J.: Application of fractional calculus in the control of heat systems. J. Adv. Comput. Intell. Intell. Inform. **11**(9), 1086–1091 (2007)
27. Machado, J.: Analysis and design of fractional-order digital control systems. SAMS **27**, 107–122 (1997)
28. Liu, L., Pann, F., Xue, D.: Variable-order fuzzy fractional PID controller. ISA Trans. **55**, 227–233 (2015)
29. Shah, P., Agashe, S.: Review of fractional PID controller. Mechatronics **38**, 29–41 (2016)
30. El-Khazali, R.: Fractional-order PIλDμ controller design. Comput. Math. Appl. **66**, 639–646 (2013)
31. Owolabi, K.M.: Robust and adaptive techniques for numerical simulation of nonlinear partial differential equations of fractional order. Commun. Nonlinear SCi Numer. Simul. **44**, 304–317 (2017)
32. Hélie, T.: Simulation of fractional-order low-pass filters. IEEE/ACM Trans. Audio Speech Lang. Process. **22**(11), 1636–1647 (2014)

33. Hwang, C., Leu, J.-F., Tsay, S.-Y.: A note on time-domain simulation of feedback fractional-order systems. IEEE Trans. Autom. Control **47**(4), 625–631 (2002)
34. Podlubny, I.: Fractional Differential Equations. Academic Press, USA (1998)
35. Miller, K.S., Ross, B.: An Introduction to the Fractional Calculus and Fractional Differential Equations. Wiley, USA (1993)
36. Dorcak, L.: Numerical Models for the Simulation of the Fractional-Order Control Systems, UEF-04-94,Kosice, Technical Report, November, (1994)
37. Barbosa, R., Machado, J.A., Silva, M.: Time domain design of fractional differintegrators using least-squares. Signal Process. **86**(10), 2567–2581 (2006)
38. Chen, Y.Q., Vinagre, B., Podlubny, I.: Continued fraction expansion to discretize fractional order derivatives—an expository review. Nonlinear Dyn. **38**(1–4), 155–170 (2004)
39. Vinagre Blas, M., Chen, Y., Petráš, I.: Two direct Tustin discretization methods for fractional-order differentiator/integrator. Frankl. Inst. **340**(5), 349–362 (2003)
40. Jacobs, B.A.: A new Grünwald-Letnikov derivative derived from a second-order scheme. Abstr. Appl. Anal. **2015**, Article ID 952057, 9 pages (2015). https://doi.org/10.1155/2015/952057
41. Scherer, R., Kalla, S.L., Tang, Y., Huang, J.: The Grünwald-Letnikov method for fractional differential equations. Comput. Math. Appl. **62**, 902–917 (2011)
42. Liua, H., Lia, S., Cao, J., Li, G., Alsaedi, A., Alsaadi, F.E.: Adaptive fuzzy prescribed performance controller design for a class of uncertain fractional-order nonlinear systems with external disturbances. Neurocomputing (In Press)
43. Bigdeli, N.: The design of a non-minimal state space fractional-order predictive functional controller for fractional systems of arbitrary order. J. Process Control **29**, 45–56 (2015)
44. Cordón, O., Herrera, F.: A three-stage evolutionary process for learning descriptive and approximate fuzzy-logic-controller knowledge bases from examples. Int. J. Approximate Reasoning **17**(4), 369–407 (1997)
45. Takagi, T., Sugeno, M.: Fuzzy identification of systems and its applications to modeling and control. IEEE Trans. Syst. Man Cybern. **SMC-15**, 116–132 (1985)
46. Mamdani, E., Assilian, S.: An experiment in linguistic synthesis with a fuzzy logic controller. Int. J. Man Mach. Stud. **7**, 1–13 (1975)
47. Xu, J.-X., Li, C., Hang, C.C.: Tuning of fuzzy PI controllers based on gain/phase margin specifications and ITAE index. ISA Trans. **35**(1), 79–91 (1996)

Chapter 4
The Metaheuristic Algorithm of the Locust-Search

Metaheuristic is a set of soft computing techniques which considers the design of intelligent search algorithms based on the analysis of several natural and social phenomena. Many metaheuristic methods have been suggested to solve a wide range of complex optimization applications. Even though these schemes have been designed to satisfy the requirements of general optimization problems, no single method can solve all problems adequately. Consequently, an enormous amount of research has been dedicated to producing new optimization methods that attain better performance indexes. In this chapter, metaheuristic algorithm called Locust Search (LS) is presented for solving optimization tasks. The LS method considers the simulation of the behavior presented in swarms of locusts as a metaphor. In the algorithm, individuals imitate a group of locusts which operate according to the biological laws of the swarm. The algorithm defines two distinct behaviors: solitary and social. Depending on the behavior, each element is undergone to a set of evolutionary operators that emulate the distinct collective behaviors typically present in the swarm.

4.1 Introduction

The collective intelligent behavior of insect or animal groups in nature such as flocks of birds, colonies of ants, schools of fish, swarms of bees and termites have attracted the attention of researchers. The aggregative conduct of insects or animals is known as swarm behavior. Even though the single members of swarms are non-sophisticated individuals, they are able to achieve complex tasks in cooperation. The collective swarm behavior emerges from relatively simple actions or interactions among the members. Entomologists have studied this collective phenomenon to model biological swarms while engineers have applied these models as a framework for solving complex real-world problems. The discipline of artificial intelligence which is concerned with the design of intelligent multi-agent

E. Cuevas et al., *Advances in Metaheuristics Algorithms: Methods and Applications*, Studies in Computational Intelligence 775,
https://doi.org/10.1007/978-3-319-89309-9_4

algorithms by taking inspiration from the collective behavior of social insects or animals is known as swarm intelligence [1]. Swarm algorithms have several advantages such as scalability, fault tolerance, adaptation, speed, modularity, autonomy and parallelism [2].

Several swarm algorithms have been developed by a combination of deterministic rules and randomness, mimicking the behavior of insect or animal groups in nature. Such methods include the social behavior of bird flocking and fish schooling such as the Particle Swarm Optimization (PSO) algorithm [3], the cooperative behavior of bee colonies such as the Artificial Bee Colony (ABC) technique [4], the social foraging behavior of bacteria such as the Bacterial Foraging Optimization Algorithm (BFOA) [5], the simulation of the herding behavior of krill individuals such as the Krill Herd (KH) method [6], the mating behavior of firefly insects such as the Firefly (FF) method [7] the emulation of the lifestyle of cuckoo birds such as the Cuckoo Search (CS) [8], the social-spider behavior such as the Social Spider Optimization (SSO) [9], the simulation of the animal behavior in a group such as the Collective Animal Behavior [10] and the emulation of the differential evolution in species such as the Differential Evolution (DE) [11].

In particular, insect swarms and animal groups provide a rich set of metaphors for designing swarm optimization algorithms. Such methods are complex systems composed by individuals that tend to reproduce specialized behaviors [12]. However, most of swarm algorithms and other evolutionary algorithms tend to exclusively concentrate the individuals in the current best positions. Under such circumstances, these algorithms seriously limit their search capacities.

Although PSO and DE are the most popular algorithms for solving complex optimization problems, they present serious flaws such as premature convergence and difficulty to overcome local minima [13, 14]. The cause for such problems is associated to the operators that modify individual positions. In such algorithms, during their evolution, the position of each agent for the next iteration is updated yielding an attraction towards the position of the best particle seen so-far (in case of PSO) or towards other promising individuals (in case of DE). As the algorithm evolves, these behaviors cause that the entire population rapidly concentrates around the best particles, favoring the premature convergence and damaging the appropriate exploration of the search space [15, 16].

The interesting and exotic collective behavior of insects have fascinated and attracted researchers for many years. The intelligent behavior observed in these groups provides survival advantages, where insect aggregations of relatively simple and "unintelligent" individuals can accomplish very complex tasks using only limited local information and simple rules of behavior [17]. Locusts (Schistocerca gregaria) are a representative example of such collaborative insects [18]. Locust is a kind of grasshopper that can change reversibly between a solitary and a social phase, which differ considerably in behavior [19]. The two phases show many differences including both overall levels of activity and the degree to which locusts are attracted or repulsed among them [20]. In the solitary phase, locusts avoid contact each other (locust concentrations). As consequence, they distribute throughout the space, exploring sufficiently the plantation [20]. On other hand, in the social phase, locusts frantically concentrate around the elements that have already found good food sources [21]. Under such a behavior, locust attempt to efficiently find better nutrients by devastating promising areas within the plantation.

In this chapter, the Locust Search (LS) is analyzed for solving optimization tasks. The LS algorithm is based on the simulation of the behavior presented in swarms of locusts. In LS, individuals emulate a group of locusts which interact to each other based on the biological laws of the cooperative swarm. The algorithm considers two different behaviors: solitary and social. Depending on the behavior, each individual is conducted by a set of evolutionary operators which mimic the different cooperative behaviors that are typically found in the swarm. Different to most of existent swarm algorithms, in LS approach, the modeled behavior explicitly avoids the concentration of individuals in the current best positions. Such fact allows not only to emulate in a better realistic way the cooperative behavior of the locust colony, but also to incorporate a computational mechanism to avoid critical flaws commonly present in the popular PSO and DE algorithms, such as the premature convergence and the incorrect exploration–exploitation balance. In order to illustrate the proficiency and robustness of the LS approach, it is compared to other well-known evolutionary methods. The comparison examines several standard benchmark functions which are commonly considered in the literature. The results show a high performance of the LS method for searching a global optimum in several benchmark functions.

This chapter is organized as follows. In Sect. 4.2, we introduce basic biological aspects and models of the algorithm. In Sect. 4.3, the novel LS algorithm and its characteristics are both described. Section 4.4 presents the experimental results and the comparative study. Finally, in Sect. 4.5, conclusions are drawn.

4.2 Biological Fundamentals

Social insect societies are complex cooperative systems that self-organize within a set of constraints. Cooperative groups are better at manipulating and exploiting their environment, defending resources and brood, and allowing task specialization among group members [22, 23]. A social insect colony functions as an integrated unit that not only possesses the ability to operate at a distributed manner, but also to undertake enormous construction of global projects [24]. It is important to acknowledge that global order in insects can arise as a result of internal interactions among members.

Locusts are a kind of grasshoppers that exhibit two opposite behavioral phases, solitary and social (gregarious). Individuals in the solitary phase avoid contact each other (locust concentrations). As consequence, they distribute throughout the space, exploring sufficiently the plantation [20]. In contrast, locusts in the gregarious phase form several concentrations. These concentrations may contain up to 10^{10} members, cover cross-sectional areas of up to 10 km^2, and travel up to 10 km per day for a period of days or weeks as they feed causing devastating crop loss [25]. The mechanism for the switch from the solitary phase to the gregarious phase is complex, and has been a subject of significant biological inquiry. A set of factors recently has been implicated, including geometry of the vegetation landscape and the olfactory stimulus [26].

Only few works [20, 21] that mathematically model the locust behavior have been published. In such approaches, it is developed two different minimal models with the goal of reproducing the macroscopic structure and motion of a group of locusts. Since the method proposed in [20] models the behavior of each locust in the group, it is used to explain the algorithm LS in this chapter.

4.2.1 Solitary Phase

In this section, it is described the way in which the position of each locust is modified as a consequence of its behavior under the solitary phase. Considering that $\mathbf{x}_i^k$ represents the current position of the ith locust in a group of N different elements, the new position $\mathbf{x}_i^{k+1}$ is calculated by using the following model:

$$\mathbf{x}_i^{k+1} = \mathbf{x}_i^k + \Delta\mathbf{x}_i, \tag{4.1}$$

where $\Delta\mathbf{x}_i$ corresponds to the change of position experimented by $\mathbf{x}_i^k$ as a consequence of its social interaction with all the other elements in the group.

Two locusts in the solitary phase exert forces on each other according to basic biological principles of attraction and repulsion (see, e.g., [20]). Repulsion operates very strongly over a short length scale in order to avoid concentrations. Attraction is weaker, and operates over a longer length scale, providing the social force necessary for maintaining the cohesion in the group. Therefore, it is modeled the strength of these social forces using the function:

$$s(r) = F \cdot e^{-r/L} - e^{-r} \tag{4.2}$$

Here, r is a distance, F describes the strength of attraction, and L is the typical attractive length scale. We have scaled the time and space coordinates so that the repulsive strength and length scale are unity. We assume that $F < 1$ and $L > 1$ so that repulsion is stronger and shorter-scale, and attraction in weaker and longer-scale. This is typical for social organisms [21]. The social force exerted by locust j on locust i is:

$$\mathbf{s}_{ij} = s(r_{ij}) \cdot \mathbf{d}_{ij}, \tag{4.3}$$

where $r_{ij} = |\mathbf{x}_j - \mathbf{x}_i|$ is the distance between the two locusts and $\mathbf{d}_{ij} = (\mathbf{x}_j - \mathbf{x}_i)/r_{ij}$ is the unit vector pointing from $\mathbf{x}_i$ to $\mathbf{x}_j$. The total social force on each locust can be modeled as the superposition of all of the pairwise interactions:

$$\mathbf{S}_i = \sum_{\substack{j=1 \\ j \neq i}}^{N} \mathbf{s}_{ij}, \tag{4.4}$$

The change of position $\Delta\mathbf{x}_i$ is modeled as the total social force experimented by $\mathbf{x}_i^k$ as the superposition of all of the pairwise interactions. Therefore, $\Delta\mathbf{x}_i$ is defined as follows:

$$\Delta\mathbf{x}_i = \mathbf{S}_i, \tag{4.5}$$

In order to illustrate the behavioral model under the solitary phase, Fig. 4.1 presents an example. It is assumed a population of three different members (N = 3) which adopt a determined configuration in the current iteration k. As a consequence of the social forces, each element suffers an attraction or repulsion to other elements depending on the distance among them. Such forces are represented by $\mathbf{s}_{12}, \mathbf{s}_{13}, \mathbf{s}_{21}, \mathbf{s}_{23}, \mathbf{s}_{31}, \mathbf{s}_{32}$. Since $\mathbf{x}_1$ and $\mathbf{x}_2$ are too close, the social forces $\mathbf{s}_{12}$ and $\mathbf{s}_{13}$ present a repulsive nature. On the other hand, as the distances $|\mathbf{x}_1 - \mathbf{x}_3|$ and $|\mathbf{x}_2 - \mathbf{x}_3|$ are quite long, the social forces $\mathbf{s}_{13}, \mathbf{s}_{23}, \mathbf{s}_{31}$ and $\mathbf{s}_{32}$ between $\mathbf{x}_1 \leftrightarrow \mathbf{x}_3$ and $\mathbf{x}_2 \leftrightarrow \mathbf{x}_3$ are from the attractive nature. Therefore, the change of position $\Delta\mathbf{x}_1$ is computed as the resultant between $\mathbf{s}_{12}$ and $\mathbf{s}_{13}$ $(\Delta\mathbf{x}_1 = \mathbf{s}_{12} + \mathbf{s}_{13})$. The values $\Delta\mathbf{x}_2$ and $\Delta\mathbf{x}_3$ of the locusts $\mathbf{x}_1$ and $\mathbf{x}_2$ are also calculated accordingly.

In addition to the presented model [20], some studies [27–29] suggest that the social force $\mathbf{s}_{ij}$ is also affected by the dominance of the involved individuals $\mathbf{x}_i$ and

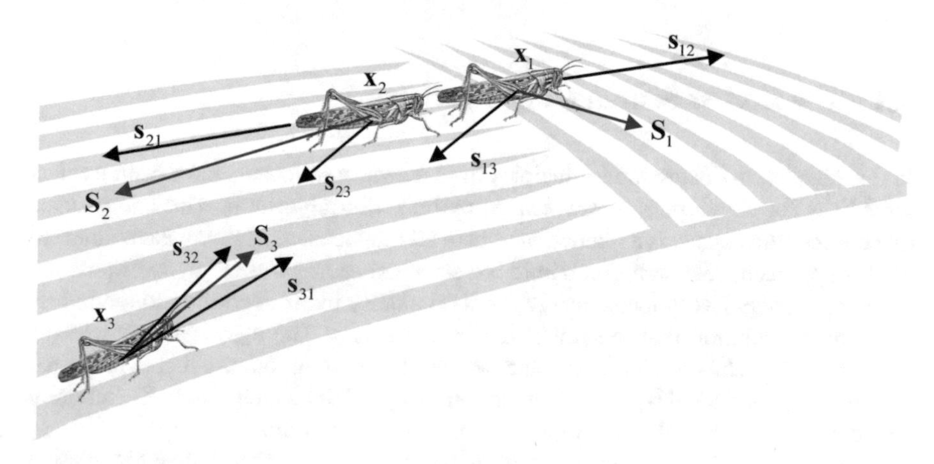

Fig. 4.1 Behavioral model under the solitary phase

$\mathbf{x}_j$ in the pairwise process. Dominance is a property that relatively qualifies the capacity of an individual to survive, in relation to other elements in a group. Dominance in locust is determined for several characteristics such as size, chemical emissions, location with regard to food sources, etc. Under such circumstances, the social force is magnified or weakened depending on the most dominant individual involved in the repulsion-attraction process.

4.2.2 *Social Phase*

In this phase, locusts frantically concentrate around the elements that have already found good food sources. Under such a behavior, locust attempt to efficiently find better nutrients by devastating promising areas within the plantation.

In order to simulate the social phase, to each locust $\mathbf{x}_i$ of the group, it is associated a food quality index Fq_i. This index reflex the quality of the food source where $\mathbf{x}_i$ is located.

Under this behavioral model, it is first ranked the N elements of the group according to their food quality indexes. Afterward, the b elements with the best food quality indexes are selected $(b \ll N)$. Considering a concentration radius R_c created around each selected element, a set of c new locusts is randomly generated inside R_c. As a result, most of the locusts will be concentrated around the best b elements. Figure 4.2 shows a simple example of behavioral model under the social phase. In the example, it is assumed a configuration of eight locust ($N = 8$), as it is illustrated in Fig. 4.2a. In the Figure, it is also presented the food quality index for each locust. A food quality index near to one indicates a better food source. Therefore, considering $b = 2$, the final configuration after the social phase, it is presented in Fig. 4.2b.

4.3 The Locust Search (LS) Algorithm

In this chapter, the behavioral principles from a swarm of locusts have been used as guidelines for developing a new swarm optimization algorithm. The LS assumes that entire search space is a plantation, where all the locusts interact to each other. In the LS approach, each solution within the search space represents a locust position in the plantation. Every locust receives a food quality index according to the fitness value of the solution that is symbolized by the locust. The algorithm implements two different behaviors: solitary and social. Depending on the behavior, each individual is conducted by a set of evolutionary operators which mimic the different cooperative behaviors that are typically found in the swarm.

From the implementation point of view, in the LS operation, a population $\mathbf{L}^k(\{\mathbf{l}_1^k, \mathbf{l}_2^k, \ldots, \mathbf{l}_N^k\})$ of N locusts (individuals) is evolved from the initial point

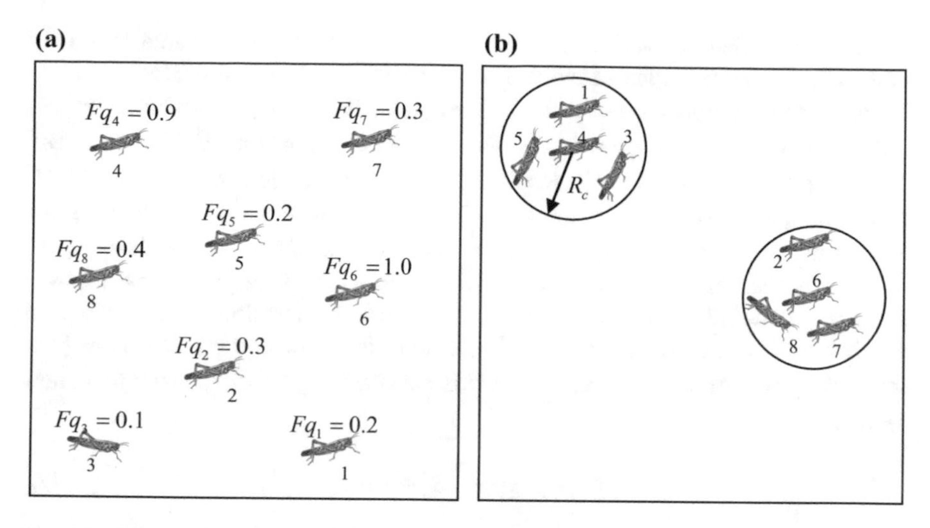

Fig. 4.2 Behavioral model under the social phase. **a** Initial configuration and food quality indexes, **b** final configuration after the operation of the social phase

($k = 0$) to a total *gen* number iterations ($k = gen$). Each locust $\mathbf{l}_i^k (i \in [1, \ldots, N])$ represents an *n*-dimensional vector $\left\{ l_{i,1}^k, l_{i,2}^k, \ldots, l_{i,n}^k \right\}$ where each dimension corresponds to a decision variable of the optimization problem to be solved. The set of decision variables constitutes the feasible search space $\mathbf{S} = \left\{ \mathbf{l}_i^k \in \mathbb{R}^n | lb_d \leq l_{i,d}^k \leq ub_d \right\}$, where lb_d and ub_d corresponds to the lower and upper bounds for the dimension *d*, respectively. The food quality index associated to each locust $\mathbf{l}_i^k$ (candidate solution) is evaluated by using an objective function $f\left(\mathbf{l}_i^k\right)$ whose final result represents the fitness value of $\mathbf{l}_i^k$. In LS, each iteration of the evolution process consists of two operators: (A) solitary and (B) social. Beginning by the solitary stage, the set of locusts is operated in order to sufficiently explore the search space. Then, during the social operation, existent solutions are refined within a determined neighborhood (exploitation).

4.3.1 Solitary Operation (A)

One of the most interesting features of the LS method is the use of the solitary operator to modify the current locust positions. Under this approach, locusts are displaced as a consequence of the social forces produced by the positional relations among the elements of the swarm. Therefore, near individuals tend to repel with each other, avoiding the concentration of elements in regions. On the other hand, distant individuals tend to attract with each other, maintaining the cohesion of the swarm. Different to the original model [20], in the proposed operator, social forces are also magnified or weakened depending on the best fitness value (the most dominant) of the individuals involved in the repulsion-attraction process.

In the solitary operation, a new position $\mathbf{p}_i(i \in [1,\ldots,N])$ is produced by perturbing the current locust position $\mathbf{l}_i^k$ with a change of position $\Delta\mathbf{l}_i(\mathbf{p}_i = \mathbf{l}_i^k + \Delta\mathbf{l}_i)$. The change of position $\Delta\mathbf{l}_i$ is the result of the social interactions experimented by $\mathbf{l}_i^k$ as a consequence of its repulsion-attraction behavioral model. Such social interactions are pairwise computed among $\mathbf{l}_i^k$ and the other $N-1$ individuals in the swarm. In the original model, social forces are calculated by using Eq. 4.3. However, in the LS method, it is modified to include the best fitness value (the most dominant) of the individuals involved in the repulsion-attraction process. Therefore, the social force exerted between $\mathbf{l}_j^k$ and $\mathbf{l}_i^k$ is calculated by using the following new model:

$$\mathbf{s}_{ij}^m = \rho(\mathbf{l}_i^k, \mathbf{l}_j^k) \cdot s(r_{ij}) \cdot \mathbf{d}_{ij} + rand(1,-1), \tag{4.6}$$

where $s(r_{ij})$ is the social force strength defined in Eq. 4.2 and $\mathbf{d}_{ij} = (\mathbf{l}_j^k - \mathbf{l}_i^k)/r_{ij}$ is the unit vector pointing from $\mathbf{l}_i^k$ to $\mathbf{l}_j^k$. Besides, *rand*(1, −1) is a number randomly generated between 1 and −1.

$\rho(\mathbf{l}_i^k, \mathbf{l}_j^k)$ is the dominance function that calculates the dominance value of the most dominant individual from $\mathbf{l}_j^k$ and $\mathbf{l}_i^k$. In order to operate $\rho(\mathbf{l}_i^k, \mathbf{l}_j^k)$, all the individuals from $\mathbf{L}^k(\{\mathbf{l}_1^k, \mathbf{l}_2^k, \ldots, \mathbf{l}_N^k\})$ are ranked according to their fitness values. The ranks are assigned so that the best individual receives the rank 0 (zero) whereas the worst individual obtains the rank $N-1$. Therefore, the function $\rho(\mathbf{l}_i^k, \mathbf{l}_j^k)$ is defined as follows:

$$\rho(\mathbf{l}_i^k, \mathbf{l}_j^k) = \begin{cases} e^{-\left(5 \cdot \text{rank}(\mathbf{l}_i^k)/N\right)} & \text{if } \text{rank}(\mathbf{l}_i^k) < \text{rank}(\mathbf{l}_j^k) \\ e^{-\left(5 \cdot \text{rank}(\mathbf{l}_j^k)/N\right)} & \text{if } \text{rank}(\mathbf{l}_i^k) > \text{rank}(\mathbf{l}_j^k) \end{cases}, \tag{4.7}$$

where the function rank(α) delivers the rank of the α-individual. According to Eq. 4.7, $\rho(\mathbf{l}_i^k, \mathbf{l}_j^k)$ gives as a result a value within the interval (1,0).

The maximum value of one is reached by $\rho(\mathbf{l}_i^k, \mathbf{l}_j^k)$ when one of the individuals $\mathbf{l}_j^k$ and $\mathbf{l}_i^k$ is the best element of the population $\mathbf{L}^k$ in terms of its fitness value. On the other hand, a value close to zero, it is obtained when both individuals $\mathbf{l}_j^k$ and $\mathbf{l}_i^k$ possess quite bad fitness values. Figure 4.3 shows the behavior of $\rho(\mathbf{l}_i^k, \mathbf{l}_j^k)$ considering 100 individuals. In the Figure, it is assumed that $\mathbf{l}_i^k$ represents one of the 99 individuals with ranks between 0 and 98 whereas $\mathbf{l}_j^k$ is fixed to the element with the worst fitness value (rank 99).

Fig. 4.3 Behavior of $\rho(\mathbf{l}_i^k, \mathbf{l}_j^k)$ considering 100 individuals

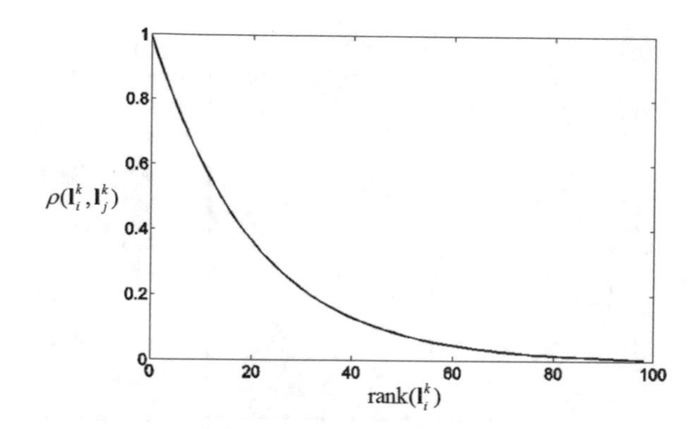

Under the incorporation of $\rho(\mathbf{l}_i^k, \mathbf{l}_j^k)$ in Eq. 4.6, social forces are magnified or weakened depending on the best fitness value (the most dominant) of the individuals involved in the repulsion-attraction process.

Finally, the total social force on each individual $\mathbf{l}_i^k$ is modeled as the superposition of all of the pairwise interactions exerted over it:

$$\mathbf{S}_i^m = \sum_{\substack{j=1 \\ j \neq i}}^{N} \mathbf{s}_{ij}^m, \tag{4.8}$$

Therefore, the change of position $\Delta\mathbf{l}_i$ is considered as the total social force experimented by $\mathbf{l}_i^k$ as the superposition of all of the pairwise interactions. Therefore, $\Delta\mathbf{l}_i$ is defined as follows:

$$\Delta\mathbf{l}_i = \mathbf{S}_i^m, \tag{4.9}$$

After calculating the new positions $\mathbf{P}(\{\mathbf{p}_1, \mathbf{p}_2, \ldots, \mathbf{p}_N\})$ of the population $\mathbf{L}^k(\{\mathbf{l}_1^k, \mathbf{l}_2^k, \ldots, \mathbf{l}_N^k\})$, the final positions $\mathbf{F}(\{\mathbf{f}_1, \mathbf{f}_2, \ldots, \mathbf{f}_N\})$ must be calculated. The idea is to admit only the changes that guarantee an improvement in the search strategy. If the fitness value of $\mathbf{p}_i(f(\mathbf{p}_i))$ is better than $\mathbf{l}_i^k(f(\mathbf{l}_i^k))$, then $\mathbf{p}_i$ is accepted as the final solution. Otherwise, $\mathbf{l}_i^k$ is retained. This procedure can be resumed by the following statement (considering a minimization problem):

$$\mathbf{f}_i = \begin{cases} \mathbf{p}_i & \text{if } f(\mathbf{p}_i) < f(\mathbf{l}_i^k) \\ \mathbf{l}_i^k & \text{otherwise} \end{cases} \tag{4.10}$$

In order to illustrate the performance of the solitary operator, Fig. 4.4 presents a simple example where the solitary operator is iteratively applied. It is assumed a population of 50 different members ($N = 50$) which adopt a concentrated configuration as initial condition (Fig. 4.4a). As a consequence of the social forces, the set

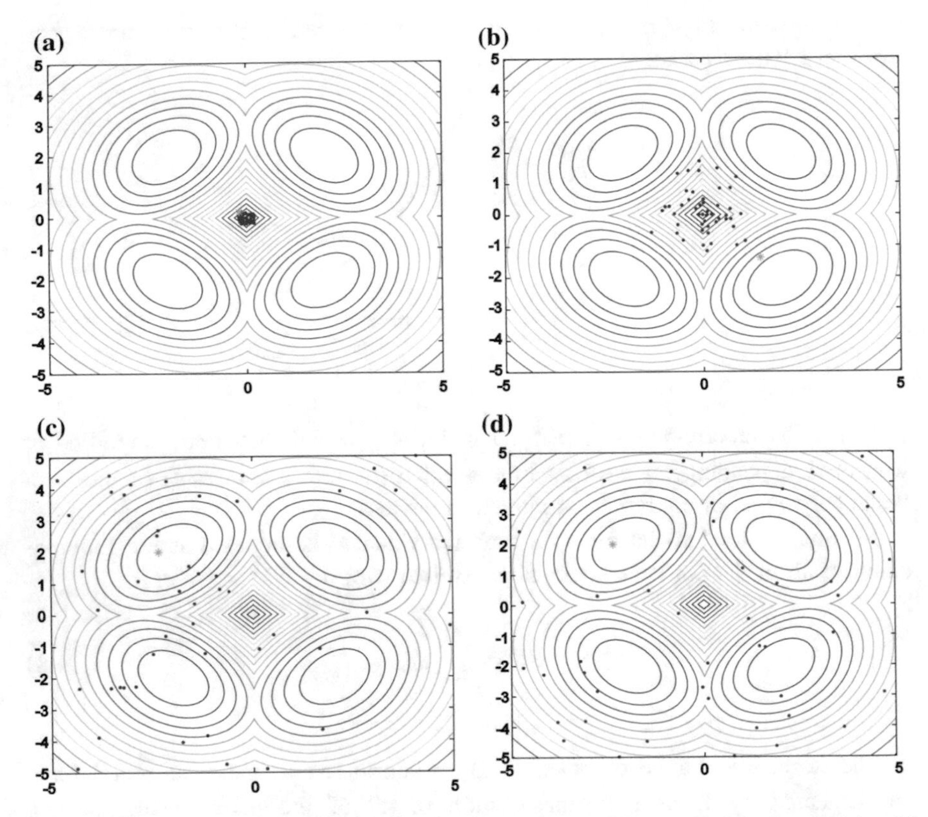

Fig. 4.4 Examples of different distributions. **a** Initial condition, **b** distribution after applying 25 operations, **c** 50 and **d** 100

of element tends to distribute through the search space. Examples of different distributions are shown in Fig. 4.4b–d after applying 25, 50 and 100 different solitary operations, respectively.

4.3.2 *Social Operation (B)*

The social procedure represents the exploitation phase of the LS algorithm. Exploitation is the process of refining existent individuals within a small neighborhood in order to improve their solution quality.

The social procedure is a selective operation which is applied only to a subset $\mathbf{E}$ of the final positions $\mathbf{F}$ (where $\mathbf{E} \subseteq \mathbf{F}$). In the operation first is necessary to sort $\mathbf{F}$ according to their fitness values and store the sorted elements in a temporal population $\mathbf{B} = \{\mathbf{b}_1, \mathbf{b}_2, \ldots, \mathbf{b}_N\}$. The elements in $\mathbf{B}$ are sorted so that the best individual receives the position $\mathbf{b}_1$ whereas the worst individual obtains the location

$\mathbf{b}_N$. Therefore, the subset $\mathbf{E}$ is integrated by only the first g locations of $\mathbf{B}$ (promising solutions). Under this operation, a subspace C_j is created around each selected particle $\mathbf{f}_j \in \mathbf{E}$. The size of C_j depends on the distance e_d which is defined as follows:

$$e_d = \frac{\sum_{q=1}^{n} \left(ub_q - lb_q \right)}{n} \cdot \beta \tag{4.11}$$

where ub_q and lb_q are the upper and lower bounds in the qth dimension, n is the number of dimensions of the optimization problem, whereas $\beta \in [0,1]$ is a tuning factor. Therefore, the limits of C_j are modeled as follows:

$$\begin{aligned} uss_j^q &= b_{j,q} + e_d \\ lss_j^q &= b_{j,q} - e_d \end{aligned} \tag{4.12}$$

where uss_j^q and lss_j^q are the upper and lower bounds of the qth dimension for the subspace C_j, respectively.

Considering the subspace C_j around each element $\mathbf{f}_j \in \mathbf{E}$, a set of h new particles $\left(\mathbf{M}_j^h = \left\{ \mathbf{m}_j^1, \mathbf{m}_j^2, \ldots, \mathbf{m}_j^h \right\} \right)$ are randomly generated inside the bounds defined by Eq. 4.12. Once the h samples are generated, the individual $\mathbf{l}_j^{k+1}$ of the next population $\mathbf{L}^{k+1}$ must be created. In order to calculate $\mathbf{l}_j^{k+1}$, the best particle $\mathbf{m}_j^{best}$, in terms of fitness value from the h samples (where $\mathbf{m}_j^{best} \in \left[\mathbf{m}_j^1, \mathbf{m}_j^2, \ldots, \mathbf{m}_j^h \right]$), is compared to $\mathbf{f}_j$. If $\mathbf{m}_j^{best}$ is better than $\mathbf{f}_j$ according to their fitness values, $\mathbf{l}_j^{k+1}$ is updated with $\mathbf{m}_j^{best}$, otherwise $\mathbf{f}_j$ is selected. The elements of $\mathbf{F}$ that have not been processed by the procedure ($\mathbf{f}_w \notin \mathbf{E}$) transfer their corresponding values to $\mathbf{L}^{k+1}$ with no change.

The social operation is used to exploit only prominent solutions. According to the propose method, inside each subspace C_j, h random samples are selected. Since the number of selected samples in each subspace is very small (typically $h < 4$), the use of this operator reduces substantially the number of fitness function evaluations.

In order to demonstrate the social operation, a numerical example has been set by applying the proposed process to a simple function. Such function considers the interval of $-3 \leq d_1, d_2 \leq 3$ whereas the function possesses one global maxima of value 8.1 at (0,1.6). Notice that d_1 and d_2 correspond to the axis coordinates (commonly x and y). For this example, it is assumed a final position population $\mathbf{F}$ of six 2-dimensional members (N = 6). Figure 4.5 shows the initial configuration of the proposed example, the black points represents the half of the particles with the best fitness values (the first three element of $\mathbf{B}$, g = 3) whereas the grey points ($\mathbf{f}_2, \mathbf{f}_4, \mathbf{f}_6 \notin \mathbf{E}$) corresponds to the remaining individuals. From Fig. 4.5, it can be seen that the social procedure is applied to all black particles ($\mathbf{f}_5 = \mathbf{b}_1, \mathbf{f}_3 = \mathbf{b}_2$ and $\mathbf{f}_1 = \mathbf{b}_3, \mathbf{f}_5, \mathbf{f}_3, \mathbf{f}_1 \in \mathbf{E}$) yielding two new random particles (h = 2), characterized by

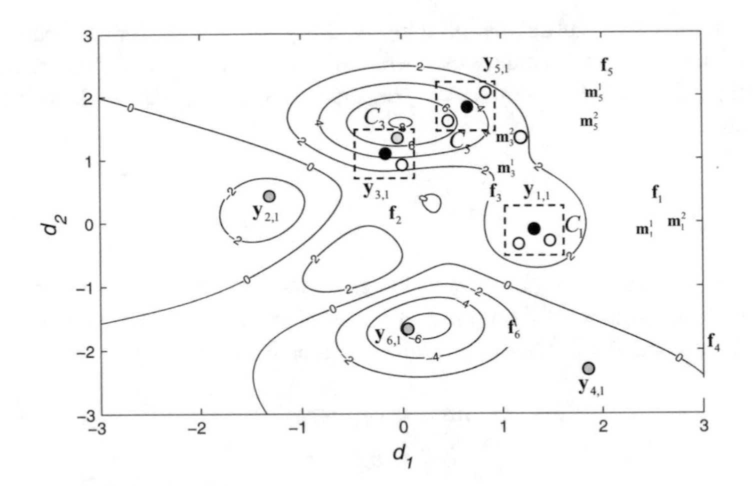

Fig. 4.5 Operation of the social procedure

the white points $\mathbf{m}_1^1, \mathbf{m}_1^2, \mathbf{m}_3^1, \mathbf{m}_3^2, \mathbf{m}_5^1$ and $\mathbf{m}_5^2$ for each black point inside of their corresponding subspaces C_1, C_3 and C_5. Considering the particle $\mathbf{f}_3$ in Fig. 4.5, the particle $\mathbf{m}_3^2$ corresponds to the best particle $\left(\mathbf{m}_3^{best}\right)$ from the two randomly generated particles (according to their fitness values) within C_3. Therefore, the particle $\mathbf{m}_3^{best}$ will substitute $\mathbf{f}_3$ in the individual $\mathbf{l}_3^{k+1}$ for the next generation, since it holds a better fitness value than $\mathbf{f}_3\left(f(\mathbf{f}_3) < f(\mathbf{m}_3^{best})\right)$.

4.3.3 Complete LS Algorithm

LS is a simple algorithm with only five adjustable parameters: the strength of attraction F, the attractive length L, number of promising solutions g, the population size N and the number of generations *gen*. The operation of LS is divided in three parts: Initialization, solitary operation and the social process. In the initialization ($k = 0$), the first population $\mathbf{L}^0\left(\{\mathbf{l}_1^0, \mathbf{l}_2^0, \ldots, \mathbf{l}_N^0\}\right)$ is produced. The values $\left\{l_{i,1}^0, l_{i,2}^0, \ldots, l_{i,n}^0\right\}$ of each individual $\mathbf{l}_i^k$ and each dimension d are randomly and uniformly distributed between the pre-specified lower initial parameter bound lb_d and the upper initial parameter bound ub_d.

$$l_{i,j}^0 = lb_d + \text{rand} \cdot (ub_d - lb_d); \quad i = 1, 2, \ldots, N; \; d = 1, 2, \ldots, n. \tag{4.13}$$

In the evolution process, the solitary (A) and social (B) operations are iteratively applied until the number of iterations $k = gen$ has been reached. The complete LS procedure is illustrated in the Algorithm 1.

Algorithm 1. Locust Search (LS) algorithm		
1:	**Input**: F, L, g, N and gen	
2:	Initialize $\mathbf{L}^0$ $(k = 0)$	
3:	**until** $(k = gen)$	
5:	$\mathbf{F} \leftarrow \text{SolitaryOperation}\left(\mathbf{L}^k\right)$	Solitary operator (3.1)
6:	$\mathbf{L}^{k+1} \leftarrow \text{SocialOperation}\left(\mathbf{L}^k, \mathbf{F}\right)$	Social operator (3.2)
8:	$k = k + 1$	
7:	**end until**	

4.3.4 Discussion About the LS Algorithm

Evolutionary algorithms (EA) have been widely employed for solving complex optimization problems. These methods are found to be more powerful than conventional methods based on formal logics or mathematical programming [30]. In an EA algorithm, search agents have to decide whether to explore unknown search positions or to exploit already tested positions in order to improve their solution quality. Pure exploration degrades the precision of the evolutionary process but increases its capacity to find new potentially solutions. On the other hand, pure exploitation allows refining existent solutions but adversely drives the process to local optimal solutions. Therefore, the ability of an EA to find a global optimal solution depends on its capacity to find a good balance between the exploitation of found-so-far elements and the exploration of the search space [31]. So far, the exploration–exploitation dilemma has been an unsolved issue within the framework of evolutionary algorithms.

Most of swarm algorithms and other evolutionary algorithms tend to exclusively concentrate the individuals in the current best positions. Under such circumstances, these algorithms seriously limit their exploration–exploitation capacities.

Different to most of existent evolutionary algorithms, in the LS approach, the modeled behavior explicitly avoids the concentration of individuals in the current best positions. Such fact allows not only to emulate in a better realistic way the cooperative behavior of the locust colony, but also to incorporate a computational mechanism to avoid critical flaws commonly present in the popular PSO and DE algorithms, such as the premature convergence and the incorrect exploration–exploitation balance.

In order to detect ellipse shapes, candidate images must be preprocessed first by the well-known Canny algorithm which yields a single-pixel edge-only image. Then, the (x_i, y_i) coordinates for each edge pixel p_i are stored inside the edge vector $P = \{p_1, p_2, \ldots, p_{N_p}\}$, with N_p being the total number of edge pixels.

4.4 Experimental Results

A comprehensive set of 13 functions, collected from Refs. [32–37], has been used to test the performance of the LS approach. Tables 4.5 and 4.6 in the Appendix present the benchmark functions used in our experimental study. Such functions are classified into two different categories: Unimodal test functions (Table 4.5) and multimodal test functions (Table 4.6). In these tables, n is the dimension of function, f_{opt} is the minimum value of the function, and $\mathbf{S}$ is a subset of R^n. The optimum location $(\mathbf{x}_{opt})$ for functions in Tables 4.5 and 4.6, are in $[0]^n$, except for f_5,f_{12},f_{13} with $\mathbf{x}_{opt}$ in $[1]^n$ and f_8 in $[420.96]^n$. A detailed description of optimum locations is given in Tables 4.5 and 4.6 of the Appendix.

We have applied the LS algorithm to 13 functions whose results have been compared to those produced by the Particle Swarm Optimization (PSO) method [3] and the Differential Evolution (DE) algorithm [11]. These are considered as the most popular algorithms for many optimization applications. In all comparisons, the population has been set to 40 (N = 40) individuals. The maximum iteration number for all functions has been set to 1000. Such stop criterion has been selected to maintain compatibility to similar works reported in the literature [34, 35].

The parameter settings for each of the algorithms in the comparison are described as follows:

1. PSO: In the algorithm, $c_1 = c_2 = 2$ while the inertia factor (ω) is decreasing linearly from 0.9 to 0.2.
2. DE: The DE/Rand/1 scheme is employed. The parameter settings follow the instructions in [11]. The crossover probability is CR = 0.9 and the weighting factor is F = 0.8.
3. In LS, F and L are set to 0.6 and L, respectively. Besides, g is fixed to 20 (N/2) whereas gen and N are configured to 1000 and 40, respectively. Once these parameters have been determined experimentally, they are kept for all experiments in this section.

Uni-modal test functions

Functions f_1 to f_7 are unimodal functions. The results for unimodal functions, over 30 runs, are reported in Table 4.1 considering the following performance indexes: the average best-so-far solution (ABS), the median of the best solution in the last iteration (MBS) and the standard deviation (SD). According to this table, LS provides better results than PSO and DE for all functions. In particular this test yields the largest difference in performance which is directly related to a better trade-off between exploration and exploitation produced by LS operators.

A non-parametric statistical significance proof known as the Wilcoxon's rank sum test for independent samples [38, 39] has been conducted with an 5% significance level, over the "average best-so-far" data of Table 4.1. Table 4.2 reports the p-values produced by Wilcoxon's test for the pair-wise comparison of the "average best so-far" of two groups. Such groups are formed by LS versus PSO and

Table 4.1 Minimization result of benchmark functions in Table 4.5 with $n = 30$

		PSO	DE	LS
f_1	ABS	1.66×10^{-1}	6.27×10^{-3}	4.55×10^{-4}
	MBS	0.23	5.85×10^{-3}	2.02×10^{-4}
	SD	3.79×10^{-1}	1.68×10^{-1}	6.98×10^{-4}
f_2	ABS	4.83×10^{-1}	2.02×10^{-1}	5.41×10^{-3}
	MBS	0.53	1.96×10^{-1}	5.15×10^{-3}
	SD	1.59×10^{-1}	0.66	1.45×10^{-2}
f_3	ABS	2.75	5.72×10^{-1}	1.61×10^{-3}
	MBS	3.16	6.38×10^{-1}	1.81×10^{-3}
	SD	1.01	0.15	1.32×10^{-3}
f_4	ABS	1.84	0.11	1.05×10^{-2}
	MBS	1.79	0.10	1.15×10^{-2}
	SD	0.87	0.05	6.63×10^{-3}
f_5	ABS	3.07	2.39	4.11×10^{-2}
	MBS	3.03	2.32	3.65×10^{-2}
	SD	0.42	0.36	2.74×10^{-3}
f_6	ABS	6.36	6.51	5.88×10^{-2}
	MBS	6.19	6.60	5.17×10^{-2}
	SD	0.74	0.87	1.67×10^{-2}
f_7	ABS	6.14	0.12	2.71×10^{-2}
	MBS	2.76	0.14	1.10×10^{-2}
	SD	0.73	0.02	1.18×10^{-2}

Maximum number of iterations = 1000

Table 4.2 p-values produced by Wilcoxon's test comparing LS versus PSO and DE over the "average best-so-far" values from Table 4.1

LS versus	PSO	DE
f_1	1.83×10^{-4}	1.73×10^{-2}
f_2	3.85×10^{-3}	1.83×10^{-4}
f_3	1.73×10^{-4}	6.23×10^{-3}
f_4	2.57×10^{-4}	5.21×10^{-3}
f_5	4.73×10^{-4}	1.83×10^{-3}
f_6	6.39×10^{-5}	2.15×10^{-3}
f_7	1.83×10^{-4}	2.21×10^{-3}

LS versus DE. As a null hypothesis, it is assumed that there is no significant difference between mean values of the two algorithms. The alternative hypothesis considers a significant difference between the "average best-so-far" values of both approaches. All p-values reported in the table are less than 0.05 (5% significance level) which is a strong evidence against the null hypothesis, indicating that the LS results are statistically significant and that it has not occurred by coincidence (i.e. due to the normal noise contained in the process).

Multimodal test functions

Multimodal functions have many local minima, being the most difficult to optimize. For multimodal functions, the final results are more important since they reflect the algorithm's ability to escape from poor local optima and locate a near-global optimum. We have done experiments on f_8 to f_{13} where the number of local minima increases exponentially as the dimension of the function increases. The dimension of these functions is set to 30. The results are averaged over 30 runs, reporting the performance indexes in Table 4.3 as follows: the average best-so-far solution (ABS), the median of the best solution in the last iteration (MBS) and the standard deviation (SD). Likewise, p-values of the Wilcoxon signed-rank test of 30 independent runs are listed in Table 4.4.

Table 4.3 Minimization result of benchmark functions in Table 4.6 with $n = 30$

		PSO	DE	LS
f_8	ABS	-6.7×10^3	-1.26×10^4	-1.26×10^4
	MBS	-5.4×10^3	-1.24×10^4	-1.23×10^4
	SD	6.3×10^2	3.7×10^2	1.1×10^2
f_9	ABS	14.8	4.01×10^{-1}	2.49×10^{-3}
	MBS	13.7	2.33×10^{-1}	3.45×10^{-3}
	SD	1.39	5.1×10^{-2}	4.8×10^{-4}
f_{10}	ABS	14.7	4.66×10^{-2}	2.15×10^{-3}
	MBS	18.3	4.69×10^{-2}	1.33×10^{-3}
	SD	1.44	1.27×10^{-2}	3.18×10^{-4}
f_{11}	ABS	12.01	1.15	1.47×10^{-4}
	MBS	12.32	0.93	3.75×10^{-4}
	SD	3.12	0.06	1.48×10^{-5}
f_{12}	ABS	6.87×10^{-1}	3.74×10^{-1}	5.58×10^{-3}
	MBS	4.66×10^{-1}	3.45×10^{-1}	5.10×10^{-3}
	SD	7.07×10^{-1}	1.55×10^{-1}	4.18×10^{-4}
f_{13}	ABS	1.87×10^{-1}	1.81×10^{-2}	1.78×10^{-2}
	MBS	1.30×10^{-1}	1.91×10^{-2}	1.75×10^{-2}
	SD	5.74×10^{-1}	1.66×10^{-2}	1.64×10^{-3}

Maximum number of iterations = 1000

Table 4.4 p-values produced by Wilcoxon's test comparing LS versus PSO and DE over the "average best-so-far" values from Table 4.3

LS versus	PSO	DE
f_8	1.83×10^{-4}	0.061
f_9	1.17×10^{-4}	2.41×10^{-4}
f_{10}	1.43×10^{-4}	3.12×10^{-3}
f_{11}	6.25×10^{-4}	1.14×10^{-3}
f_{12}	2.34×10^{-5}	7.15×10^{-4}
f_{13}	4.73×10^{-4}	0.071

For f_9,f_{10},f_{11} and f_{12}, LS yields a much better solution than the others. However, for functions f_8 and f_{13}, LS produces similar results to DE. The Wilcoxon rank test results, presented in Table 4.4, show that LS performed better than PSO and DE considering the four problems f_9–f_{12}, whereas, from a statistical viewpoint, there is not difference in results between LS and DE for f_8 and f_{13}.

4.5 Conclusions

In this chapter, the Locust Search (LS) has been analyzed for solving optimization tasks. The LS algorithm is based on the simulation of the behavior presented in swarms of locusts. In the LS algorithm, individuals emulate a group of locusts which interact to each other based on the biological laws of the cooperative swarm. The algorithm considers two different behaviors: solitary and social. Depending on the behavior, each individual is conducted by a set of evolutionary operators which mimic the different cooperative behaviors that are typically found in the swarm.

Different to most of existent evolutionary algorithms, in the LS approach, the modeled behavior explicitly avoids the concentration of individuals in the current best positions. Such fact allows not only to emulate in a better realistic way the cooperative behavior of the locust colony, but also to incorporate a computational mechanism to avoid critical flaws commonly present in the popular PSO and DE algorithms, such as the premature convergence and the incorrect exploration–exploitation balance.

LS has been experimentally tested considering a suite of 13 benchmark functions. The performance of LS has been also compared to the following algorithms: the Particle Swarm Optimization method (PSO) [3], and the Differential Evolution (DE) algorithm [11]. Results have confirmed an acceptable performance of the LS method in terms of the solution quality for all tested benchmark functions.

The LS remarkable performance is associated with two different reasons: (i) the solitary operator allows a better particle distribution in the search space, increasing the algorithm's ability to find the global optima; and (ii) the use of the social operation, provides of a simple exploitation operator that intensifies the capacity of finding better solutions during the evolution process.

Appendix: List of Benchmark Functions

In Table 4.5, n is the dimension of function, f_{opt} is the minimum value of the function, and $\mathbf{S}$ is a subset of R^n. The optimum location $\left(\mathbf{x}_{opt}\right)$ for functions in Table 4.5 is in $[0]^n$, except for f_5 with $\mathbf{x}_{opt}$ in $[1]^n$.

The optimum location $\left(\mathbf{x}_{opt}\right)$ for functions in Table 4.6, are in $[0]^n$, except for f_8 in $[420.96]^n$ and f_{12}–f_{13} in $[1]^n$.

Table 4.5 Unimodal test functions

Test function	S	f_{opt}
$f_1(\mathbf{x}) = \sum_{i=1}^{n} x_i^2$	$[-100,100]^n$	0
$f_2(\mathbf{x}) = \sum_{i=1}^{n} \lvert x_i\rvert + \prod_{i=1}^{n} \lvert x_i\rvert$	$[-10,10]^n$	0
$f_3(\mathbf{x}) = \sum_{i=1}^{n} \left(\sum_{j=1}^{i} x_j\right)^2$	$[-100,100]^n$	0
$f_4(\mathbf{x}) = \max_i\{\lvert x_i\rvert, 1 \le i \le n\}$	$[-100,100]^n$	0
$f_5(\mathbf{x}) = \sum_{i=1}^{n-1} \left[100\left(x_{i+1} - x_i^2\right)^2 + (x_i - 1)^2\right]$	$[-30,30]^n$	0
$f_6(\mathbf{x}) = \sum_{i=1}^{n} (x_i + 0.5)^2$	$[-100,100]^n$	0
$f_7(\mathbf{x}) = \sum_{i=1}^{n} ix_i^4 + rand(0,1)$	$[-1.28,1.28]^n$	0

Table 4.6 Multimodal test functions

Test function	S	f_{opt}
$f_8(\mathbf{x}) = \sum_{i=1}^{n} -x_i \sin\left(\sqrt{\lvert x_i\rvert}\right)$	$[-500,500]^n$	$-418.98*n$
$f_9(\mathbf{x}) = \sum_{i=1}^{n} \left[x_i^2 - 10\cos(2\pi x_i) + 10\right]$	$[-5.12,5.12]^n$	0
$f_{10}(\mathbf{x}) = -20\exp\left(-0.2\sqrt{\frac{1}{n}\sum_{i=1}^{n} x_i^2}\right) - \exp\left(\frac{1}{n}\sum_{i=1}^{n}\cos(2\pi x_i)\right) + 20$	$[-32,32]^n$	0
$f_{11}(\mathbf{x}) = \frac{1}{4000}\sum_{i=1}^{n} x_i^2 - \prod_{i=1}^{n}\cos\left(\frac{x_i}{\sqrt{i}}\right) + 1$	$[-600,600]^n$	0
$f_{12}(\mathbf{x}) = \frac{\pi}{n}\left\{10\sin(\pi y_1) + \sum_{i=1}^{n-1}(y_i - 1)^2\left[1 + 10\sin^2(\pi y_{i+1})\right] + (y_n - 1)^2\right\} + \sum_{i=1}^{n} u(x_i, 10, 100, 4)$ $y_i = 1 + \frac{x_i + 1}{4}$ $u(x_i, a, k, m) = \begin{cases} k(x_i - a)^m & x_i > a \\ 0 & -a < x_i < a \\ k(-x_i - a)^m & x_i < -a \end{cases}$	$[-50,50]^n$	0
$f_{13}(\mathbf{x}) = 0.1\left\{\sin^2(3\pi x_1) + \sum_{i=1}^{n}(x_i - 1)^2\left[1 + \sin^2(3\pi x_i + 1)\right] + (x_n - 1)^2\left[1 + \sin^2(2\pi x_n)\right]\right\} + \sum_{i=1}^{n} u(x_i, 5, 100, 4)$	$[-50,50]^n$	0

References

1. Bonabeau, E., Dorigo, M., Theraulaz, G.: Swarm Intelligence: From Natural to Artificial Systems. Oxford University Press Inc, New York (1999)
2. Kassabalidis, I., El-Sharkawi, M.A., Marks II, R.J., Arabshahi, P., Gray, A.A.: Swarm intelligence for routing in communication networks. In: Global Telecommunications Conference, GLOBECOM '01, IEEE, vol. 6, pp. 3613–3617 (2001)
3. Kennedy, J., Eberhart, R.: Particle swarm optimization. In: Proceedings of the 1995 IEEE International Conference on Neural Networks, vol. 4, pp. 1942–1948, December 1995
4. Karaboga, D.: An idea based on honey bee swarm for numerical optimization. Technical Report-TR06. Engineering Faculty, Computer Engineering Department, Erciyes University (2005)

5. Passino, K.M.: Biomimicry of bacterial foraging for distributed optimization and control. IEEE Control Syst. Mag. **22**(3), 52–67 (2002)
6. Hossein, A., Hossein-Alavi, A.: Krill herd: a new bio-inspired optimization algorithm. Commun. Nonlinear Sci. Numer. Simul. **17**, 4831–4845 (2012)
7. Yang, X.S.: Engineering Optimization: An Introduction with Metaheuristic Applications. Wiley, USA (2010)
8. Yang, X.S., Deb, S.: Proceedings of World Congress on Nature & Biologically Inspired Computed, pp. 210–214. IEEE Publications, India (2009)
9. Cuevas, E., Cienfuegos, M., Zaldívar, D., Pérez-Cisneros, M.: A swarm optimization algorithm inspired in the behavior of the social-spider. Expert Syst. Appl. **40**(16), 6374–6384 (2013)
10. Cuevas, E., González, M., Zaldivar, D., Pérez-Cisneros, M., García, G.: An algorithm for global optimization inspired by collective animal behaviour. Discrete Dyn. Nat. Soc. art. no. 638275 (2012)
11. Storn, R., Price, K.: Differential evolution—a simple and efficient adaptive scheme for global optimisation over continuous spaces. Technical Report TR-95–012. ICSI, Berkeley, CA (1995)
12. Bonabeau, E.: Social insect colonies as complex adaptive systems. Ecosystems **1**, 437–443 (1998)
13. Wang, Y., Li, B., Weise, T., Wang, J., Yuan, B., Tian, Q.: Self-adaptive learning based particle swarm optimization. Inf. Sci. **181**(20), 4515–4538 (2011)
14. Tvrdík, J.: Adaptation in differential evolution: a numerical comparison. Appl. Soft Comput. **9** (3), 1149–1155 (2009)
15. Wang, H., Sun, H., Li, C., Rahnamayan, S., Jeng-shyang, P.: Diversity enhanced particle swarm optimization with neighborhood. Inf. Sci. **223**, 119–135 (2013)
16. Gong, W., Fialho, Á., Cai, Z., Li, H.: Adaptive strategy selection in differential evolution for numerical optimization: an empirical study. Inf. Sci. **181**(24), 5364–5386 (2011)
17. Gordon, D.: The organization of work in social insect colonies. Complexity **8**(1), 43–46 (2003)
18. Kizaki, S., Katori, M.: A Stochastic lattice model for locust outbreak. Phys. A **266**, 339–342 (1999)
19. Rogers, S.M., Cullen, D.A., Anstey, M.L., Burrows, M., Dodgson, T., Matheson, T., Ott, S. R., Stettin, K., Sword, G.A., Despland, E., Simpson, S.J.: Rapid behavioural gregarization in the desert locust, Schistocerca gregaria entails synchronous changes in both activity and attraction to conspecifics. J. Insect Physiol. **65**, 9–26 (2014)
20. Topaz, C.M., Bernoff, A.J., Logan, S., Toolson, W.: A model for rolling swarms of locusts. Eur. Phys. J. Special Topics **157**, 93–109 (2008)
21. Topaz, C.M., D'Orsogna, M.R., Edelstein-Keshet, L., Bernoff, A.J.: Locust dynamics: behavioral phase change and swarming. PLoS Comput. Biol. **8**(8), 1–11
22. Oster, G., Wilson, E.: Caste and Ecology in the Social Insects. N.J. Princeton University Press, Princeton (1978)
23. Hölldobler, B., Wilson, E.O.: Journey to the Ants: A Story of Scientific Exploration (1994). ISBN 0-674-48525-4
24. Hölldobler, B., Wilson, E.O.: The Ants. Harvard University Press, USA (1990). ISBN 0-674-04075-9
25. Tanaka, S., Nishide, Y.: Behavioral phase shift in nymphs of the desert locust, Schistocerca gregaria: special attention to attraction/avoidance behaviors and the role of serotonin. J. Insect Physiol. **59**, 101–112 (2013)
26. Gaten, E., Huston, S.J., Dowse, H.B., Matheson, T.: Solitary and gregarious locusts differ in circadian rhythmicity of a visual output neuron. J. Biol. Rhythms **27**(3), 196–205 (2012)
27. Benaragama, I., Gray, J.R.: Responses of a pair of flying locusts to lateral looming visual stimuli. J. Comp. Physiol. A. **200**(8), 723–738 (2014)
28. Sergeev, M.G.: Distribution patterns of grasshoppers and their kin in the boreal zone. Psyche J. Entomol. **2011**, 9 pages, Article ID 324130 (2011)

29. Ely, S.O., Njagi, P.G.N., Bashir, M.O., El-Amin, S.E.-T., Hassanali1, A.: Diel behavioral activity patterns in adult solitarious desert locust, Schistocerca gregaria (Forskål). Psyche J. Entomol. **2011**, Article ID 459315, 9 (2011)
30. Yang, X.-S.: Nature-Inspired Metaheuristic Algorithms. Luniver Press, Beckington (2008)
31. Cuevas, E., Echavarría, A., Ramírez-Ortegón, M.A.: An optimization algorithm inspired by the states of matter that improves the balance between exploration and exploitation. Appl. Intell. **40**(2), 256–272 (2014)
32. Ali, M.M., Khompatraporn, C., Zabinsky, Z.B.: A numerical evaluation of several stochastic algorithms on selected continuous global optimization test problems. J. Global Optim. **31**(4), 635–672 (2005)
33. Chelouah, R., Siarry, P.: A continuous genetic algorithm designed for the global optimization of multimodal functions. J. Heuristics **6**(2), 191–213 (2000)
34. Herrera, F., Lozano, M., Sánchez, A.M.: A taxonomy for the crossover operator for real-coded genetic algorithms: an experimental study. Int. J. Intell. Syst. **18**(3), 309–338 (2003)
35. Laguna, M., Martí, R.: Experimental testing of advanced scatter search designs for global optimization of multimodal functions. J. Global Optim. **33**(2), 235–255 (2005)
36. Lozano, M., Herrera, F., Krasnogor, N., Molina, D.: Real-coded memetic algorithms with crossover hill-climbing. Evol. Comput. **12**(3), 273–302 (2004)
37. Moré, J.J., Garbow, B.S., Hillstrom, K.E.: Testing unconstrained optimization software. ACM Trans. Math. Softw. **7**(1), 17–41 (1981)
38. Wilcoxon, F.: Individual comparisons by ranking methods. Biometrics **1**, 80–83 (1945)
39. Garcia, S., Molina, D., Lozano, M., Herrera, F.: A study on the use of non-parametric tests for analyzing the evolutionary algorithms' behaviour: a case study on the CEC '2005, Special session on real parameter optimization. J. Heurist (2008). https://doi.org/10.1007/s10732-008-9080-4

Chapter 5
Identification of Fractional Chaotic Systems by Using the Locust Search Algorithm

Parameter estimation of fractional chaotic models has drawn the interests of different research communities due to its multiple applications. In the estimation process, the task is converted into a multi-dimensional optimization problem. Under this approach, the fractional elements, as well as functional factors of the chaotic model are assumed as decision variables. Many methods based on metaheuristic concepts have been successfully employed to identify the parameters of fractional-order chaotic models. Nevertheless, most of them present a significant weakness, they usually reach sub-optimal solutions as a consequence of an incorrect balance between exploration and exploitation in their search procedures. This chapter analyses the way in which metaheuristic algorithms can be applied for parameter identification of chaotic systems. To identify the parameters, the chapter explores the use of the metaheuristic method called Locust Search (LS) which is based on the operation of swarms of locusts. Contrary to the most of existent metaheuristic algorithms, it explicitly discourages the clustering of individuals in the promising positions, eliminating the significant defects such as the premature convergence to sub-optimal solutions and the limited exploration–exploitation balance.

5.1 Introduction

A fractional order model is a system that is characterized by a fractional differential equation containing derivatives of non-integer order. Several engineering problems, such as transmission lines [1], electrical circuits [2] and control systems [3], can be more accurately described by fractional differential equations than integer order schemes. For this reason, in the last decade, the fractional order systems [4–8] have attracted the interests of several research communities.

System identification is a practical way to model a fractional order system. However, because the mathematical interpretation of fractional calculus is lightly

E. Cuevas et al., *Advances in Metaheuristics Algorithms: Methods and Applications*, Studies in Computational Intelligence 775,
https://doi.org/10.1007/978-3-319-89309-9_5

distinct to integer calculus, it is difficult to model real fractional order systems directly based on analytic mechanisms [9]. For classical integer order system, once the maximum order of the system has been defined, the parameters of the model can be identified directly. However, for a fractional order system, because identification requires the choice of the fractional order of the operators, and the systematic parameters, the identification process of such systems is more complex than that of the integer order models [10]. Under such conditions, most of the classical identification methods cannot directly applied to identification of a fractional order systems [11].

The problem of estimating the parameters of fractional order systems has been commonly solved through the use of deterministic methods such as non-linear optimization techniques [12], input output frequency contents [13] or operational matrix [14]. These methods have been exhaustively analyzed and represents the most consolidated available tools. The interested reader in such approaches can be referred to [15] for a recent survey on the state-of-the-art.

As an alternative to classical techniques, the problem of identification in fractional order systems has also been handled through evolutionary methods. In general, they have demonstrated, under several circumstances, to deliver better results than those based on deterministic approaches in terms of accuracy and robustness [16]. Under these methods, an individual is represented by a candidate model. Just as the evolution process unfolds, a set of evolutionary operators are applied in order to produce better individuals. The quality of each candidate solution is evaluated through an objective function whose final result represents the affinity between the estimated model and the actual one. Some examples of these approaches used in the identification of fractional order systems involve methods such as Genetic Algorithms (GA) [17], Artificial Bee Colony (ABC) [18], Differential Evolution (DE) [19] and Particle Swarm Optimization (PSO) [20]. Although these algorithms present interesting results, they have an important limitation: They frequently obtain sub-optimal solutions as a consequence of the limited balance between exploration and exploitation in their search strategies. This limitation is associated to their evolutionary operators employed to modify the individual positions. In such algorithms, during their operation, the position of each individual for the next iteration is updated producing an attraction towards the position of the best particle seen so-far or towards other promising individuals. Therefore, as the algorithm evolves, such behaviors cause that the entire population rapidly concentrates around the best particles, favoring the premature convergence and damaging the appropriate exploration of the search space [21, 22].

This chapter presents an algorithm for parameter identification of fractional-order chaotic systems. In order to determine the parameters, the method uses a novel evolutionary method called Locust Search (LS) [23–25] which is based on the behavior of swarms of locusts. In the LS algorithm, individuals emulate a group of locusts which interact to each other based on the biological laws of the cooperative swarm. The algorithm considers two different behaviors: solitary and social. Depending on the behavior, each individual is conducted by a set of evolutionary operators which mimics different cooperative conducts that are typically found in the swarm. Different to most of existent evolutionary algorithms, the behavioral model in the LS approach explicitly avoids the concentration of individuals in the current best positions. Such fact allows to avoid critical flaws such as the premature convergence to sub-optimal solutions and the incorrect exploration–exploitation balance. Numerical simulations have been conducted on the fractional-Order Van der Pol oscillator to show the effectiveness of the scheme.

The chapter is organized as follows. In Sect. 5.2, the concepts of fractional calculus are introduced. Section 5.3 gives a description for the Locust Search algorithm. Section 5.4 gives a brief description of the fractional-order Van der Pol Oscillator. Section 5.5 formulates the parameter estimation problem. Section 5.6 shows the experimental results. Finally some conclusions are discussed in Sect. 5.7.

5.2 Fractional Calculus

Fractional calculus is a generalization of integration and differentiation to non-integer order fundamental operator. The differential-integral operator, denoted by ${}_aD_t^q$ takes both the fractional derivative and the fractional integral in a single expression which is defined as:

$$
{}_aD_t^q = \begin{cases} \frac{d^q}{dt^q}, & q > 0, \\ 1, & q = 0, \\ \int_a^t (d\tau)^q, & q < 0. \end{cases} \tag{5.1}
$$

where a and t represents the operation bounds whereas $q \in \Re$. The commonly used definitions for fractional derivatives are the Grünwald-Letnikov, Riemann-Liouville [7] and Caputo [26]. According to the Grünwald-Letnikov approximation, the fractional-order derivative of order q is defined as follows:

$$
D_t^q f(t) = \lim_{h \to 0} \frac{1}{h^q} \sum_{j=0}^{\infty} (-1)^j \binom{q}{j} f(t - jh) \tag{5.2}
$$

In the numerical calculation of fractional-order derivatives, the explicit numerical approximation of the qth derivative at the points $kh, (k = 1, 2, \ldots)$ maintains the following form [27]:

$$
{}_{(k - L_m/h)}D_{t_k}^q f(t) \approx h^{-q} \sum_{j=0}^{k} (-1)^j \binom{q}{j} f(t_k - j) \tag{5.3}
$$

where L_m is the memory length $t_k = kh$, h, is the time step and $(-1)^j \binom{q}{j}$ are the binomial coefficients. For their calculation we can use the following expression:

$$c_0^{(q)} = 1, \quad c_j^{(q)} = \left(1 - \frac{1+q}{j}\right) c_{j-1}^{(q)} \tag{5.4}$$

Then, the general numerical solution of the fractional differential equation is defined as follows:

$$y(t_k) = f(y(t_k), t_k)h^q - \sum_{j=1}^{k} c_j^{(q)} y(t_{k-j}) \tag{5.5}$$

5.3 Locust Search (LS) Algorithm

In the operation of LS [23], a population $\mathbf{L}^k(\{\mathbf{l}_1^k, \mathbf{l}_2^k, \ldots, \mathbf{l}_N^k\})$ of N locusts (individuals) is processed from the initial stage ($k = 0$) to a total *gen* number iterations ($k = gen$). Each individual $\mathbf{l}_i^k$ ($i \in [1, \ldots, N]$) symbolizes an n-dimensional vector $\left\{l_{i,1}^k, l_{i,2}^k, \ldots, l_{i,n}^k\right\}$ where each dimension represents a domain variable of the optimization problem to be solved. The set of variables represents the valid search space $\mathbf{S} = \left\{\mathbf{l}_i^k \in \mathbb{R}^n \middle| lb_d \leq l_{i,d}^k \leq ub_d\right\}$, where lb_d and ub_d represents the lower and upper bounds for the d dimension, respectively. The quality of each element $\mathbf{l}_i^k$ (candidate solution) is evaluated by using the objective function $f\left(\mathbf{l}_i^k\right)$. In LS, at each iteration consists of two operators: (A) solitary and (B) social.

5.3.1 *Solitary Operation (A)*

In the solitary operation, a new location $\mathbf{p}_i (i \in [1, \ldots, N])$ is generated by modifying the current element location $\mathbf{l}_i^k$ with a change of position $\Delta\mathbf{l}_i\left(\mathbf{p}_i = \mathbf{l}_i^k + \Delta\mathbf{l}_i\right)$. $\Delta\mathbf{l}_i$ is the result of the individual interactions experimented by $\mathbf{l}_i^k$ as a consequence of its biological behavior. Such interactions are pairwise computed among $\mathbf{l}_i^k$ and the other $N - 1$ individuals in the swarm. Therefore, the final force exerted between $\mathbf{l}_j^k$ and $\mathbf{l}_i^k$ is computed by considering the following model:

$$\mathbf{s}_{ij}^m = \rho(\mathbf{l}_i^k, \mathbf{l}_j^k) \cdot s(r_{ij}) \cdot \mathbf{d}_{ij} + rand(1, -1) \tag{5.6}$$

where $\mathbf{d}_{ij} = (\mathbf{l}_j^k - \mathbf{l}_i^k)/r_{ij}$ is the unit-vector, pointing from $\mathbf{l}_i^k$ to $\mathbf{l}_j^k$. Furthermore, *rand* (1, −1) is an number randomly produced between 1 and −1. The factor $s(r_{ij})$ represents the social relation between $\mathbf{l}_j^k$ and $\mathbf{l}_i^k$, which is calculated as follows:

$$s(r_{ij}) = F \cdot e^{-r_{ij}/L} - e^{-r_{ij}} \tag{5.7}$$

Here, r_{ij} is the distance between $\mathbf{l}_j^k$ and $\mathbf{l}_i^k$, F represents the strength of attraction whereas L is the attractive length factor. It is assumed that $F < 1$ and $L > 1$ so that repulsion is stronger in a shorter-scale, while attraction is applied in a weaker and longer-scale. $\rho(\mathbf{l}_i^k, \mathbf{l}_j^k)$ is a function that calculates the dominance value of the best element between $\mathbf{l}_j^k$ and $\mathbf{l}_i^k$. In order to operate $\rho(\mathbf{l}_i^k, \mathbf{l}_j^k)$, all the individuals from $\mathbf{L}^k(\{\mathbf{l}_1^k, \mathbf{l}_2^k, \ldots, \mathbf{l}_N^k\})$ are arranged in terms of their fitness values. Therefore, a rank is assigned to each element, so that the best individual obtains the rank 0 (zero) whereas the worst individual receives the rank $N - 1$. Under such conditions, the function $\rho(\mathbf{l}_i^k, \mathbf{l}_j^k)$ is defined as follows:

$$\rho(\mathbf{l}_i^k, \mathbf{l}_j^k) = \begin{cases} e^{-\left(5 \cdot \mathrm{rank}(\mathbf{l}_i^k)/N\right)} & \text{if } \mathrm{rank}(\mathbf{l}_i^k) < \mathrm{rank}(\mathbf{l}_j^k) \\ e^{-\left(5 \cdot \mathrm{rank}(\mathbf{l}_j^k)/N\right)} & \text{if } \mathrm{rank}(\mathbf{l}_i^k) > \mathrm{rank}(\mathbf{l}_j^k) \end{cases} \tag{5.8}$$

where rank (α) delivers the rank of the α-element. According to Eq. (5.8), $\rho(\mathbf{l}_i^k, \mathbf{l}_j^k)$ gives a value within [0,1]. Figure 5.1 shows the behavior of $\rho(\mathbf{l}_i^k, \mathbf{l}_j^k)$ considering 100 elements. In the figure, it is assumed that $\mathbf{l}_i^k$ represents one of the 99 individuals with ranks among 0 and 98 whereas $\mathbf{l}_j^k$ is fixed to the worst individual (rank 99).

Then, the resultant force $\mathbf{S}_i^m$ on each element $\mathbf{l}_i^k$ is computed as the superposition of all of the pairwise interactions exerted on it:

$$\mathbf{S}_i^m = \sum_{\substack{j=1 \\ j \neq i}}^{N} \mathbf{s}_{ij}^m \tag{5.9}$$

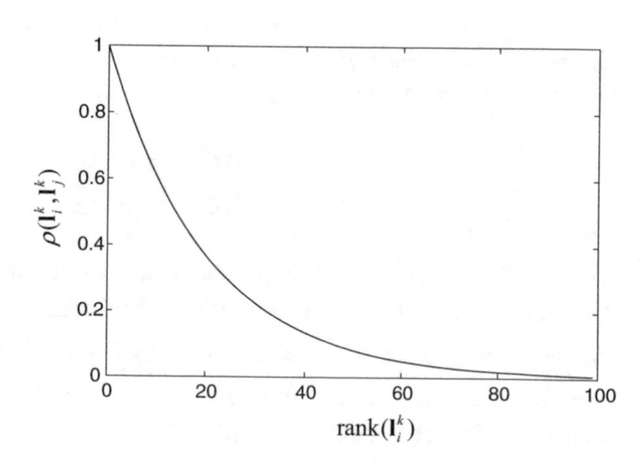

Fig. 5.1 Behavior of $\rho(\mathbf{l}_i^k, \mathbf{l}_j^k)$ considering 100 individuals

Finally, $\Delta\mathbf{l}_i$ is assumed similar to the social force experimented by $\mathbf{l}_i^k$ as the superposition of all of the pairwise reciprocal forces. Consequently, $\Delta\mathbf{l}_i$ is represented as follows:

$$\Delta\mathbf{l}_i = \mathbf{S}_i^m \tag{5.10}$$

After calculating the new locations $\mathbf{P}(\{\mathbf{p}_1, \mathbf{p}_2, \ldots, \mathbf{p}_N\})$ of the population $\mathbf{L}^k(\{\mathbf{l}_1^k, \mathbf{l}_2^k, \ldots, \mathbf{l}_N^k\})$, the final locations $\mathbf{F}(\{\mathbf{f}_1, \mathbf{f}_2, \ldots, \mathbf{f}_N\})$ must be computed. This procedure can be summarized by the following formulation (in terms of a minimization problem):

$$\mathbf{f}_i = \begin{cases} \mathbf{p}_i & \text{if } f(\mathbf{p}_i) < f(\mathbf{l}_i^k) \\ \mathbf{l}_i^k & \text{otherwise} \end{cases} \tag{5.11}$$

5.3.2 *Social Operation (B)*

The social operation is a discriminating operation which considers only to a subset $\mathbf{E}$ of the final positions $\mathbf{F}$ (where $\mathbf{E} \subseteq \mathbf{F}$). In the process first is necessary to order $\mathbf{F}$ in terms of their fitness values and collect the individuals in a temporal population $\mathbf{B} = \{\mathbf{b}_1, \mathbf{b}_2, \ldots, \mathbf{b}_N\}$. The individuals of $\mathbf{B}$ are arranged so that the best element is located in the first position $\mathbf{b}_1\{b_{1,1}, b_{1,2}, \ldots, b_{1,n}\}$ whereas the worst individual is situated in the last location $\mathbf{b}_N$. Under such conditions, $\mathbf{E}$ is composed by the first g position of $\mathbf{B}$ (the best elements). Then, a subspace C_j is defined around each selected element $\mathbf{f}_j \in \mathbf{E}$. The size of C_j depends on the distance e_d which is determined as follows:

$$e_d = \frac{\sum_{q=1}^{n} \left(ub_q - lb_q\right)}{n} \cdot \beta \tag{5.12}$$

where ub_q and lb_q are the upper and lower limits of the qth dimension, n is the number of dimensions of the optimization problem, whereas $\beta \in [0,1]$ is a tuning factor. Therefore, the bounds of C_j are modeled as follows:

$$\begin{aligned} uss_j^q &= b_{j,q} + e_d \\ lss_j^q &= b_{j,q} - e_d \end{aligned} \tag{5.13}$$

where uss_j^q and lss_j^q are the upper and lower limits of the *q-th*-dimension for the subspace C_j, respectively. Once creating the subspace C_j in the neighborhood of the element $\mathbf{f}_j \in \mathbf{E}$, a set of h new elements ($\mathbf{M}_j^h = \left\{\mathbf{m}_j^1, \mathbf{m}_j^2, \ldots, \mathbf{m}_j^h\right\}$) are randomly produced within the limits defined by Eq. 5.13. Considering the h samples, the new individual $\mathbf{l}_j^{k+1}$ of the next population $\mathbf{L}^{k+1}$ must be extracted. In order to select

$\mathbf{l}_j^{k+1}$, the best element $\mathbf{m}_j^{best}$, in terms of fitness value from the h samples (where $\mathbf{m}_j^{best} \in \left[\mathbf{m}_j^1, \mathbf{m}_j^2, \ldots, \mathbf{m}_j^h\right]$), is examined. If $\mathbf{m}_j^{best}$ is better than $\mathbf{f}_j$ according to their fitness values, $\mathbf{l}_j^{k+1}$ is updated with $\mathbf{m}_j^{best}$, otherwise the position of $\mathbf{f}_j$ is assigned to $\mathbf{l}_j^{k+1}$. The elements of $\mathbf{F}$ that have not been considered by the procedure ($\mathbf{f}_w \notin \mathbf{E}$) transport their corresponding values to $\mathbf{L}^{k+1}$ without variation. The social operation is used to exploit only favorable solutions. According to the social operation, inside each subspace C_j, h random samples are produced. Since the number of selected elements in each subspace is very small (typically $h<4$), the use of this operator cannot be considered computational expensive.

5.4 Fractional-Order van der Pol Oscillator

The Van der Pol Oscillator model has been extensively studied as a complex example of non-linear system. It provides important models for a wide range of dynamic behaviors for several engineering applications [28, 29]. The classical integer-order Van der Pol Oscillator is described by a second-order non-linear differential equation as follows:

$$\begin{bmatrix} \dot{y}_1 \\ \dot{y}_2 \end{bmatrix} = \begin{bmatrix} 0 & 1 \\ -1 & -\varepsilon(y_1^2(t) - 1) \end{bmatrix} \begin{bmatrix} y_1 \\ y_2 \end{bmatrix}, \tag{5.14}$$

where ε is a control parameter that reflects the nonlinearity degree of the system. On the other hand, the fractional-order Van der Pol Oscillator model of order q is defined by the following formulation [30]:

$$\begin{aligned} {}_0D_t^{q_1} y_1(t) &= y_2(t), \\ {}_0D_t^{q_2} y_2(t) &= -y_1(t) - \varepsilon(y_1^2(t) - 1)y_2(t). \end{aligned} \tag{5.15}$$

Considering the Grünwald-Letnikov approximation (see Eq. 5.5), the numerical solution for the fractional-order Van der Pol Oscillator is given by:

$$\begin{aligned} y_1(t_k) &= y_2(t_{k-1})h^{q_1} - \sum_{j=1}^{k} c_j^{(q_1)} y_1(t_{k-j}), \\ y_2(t_k) &= (-y_1(t_k) - \varepsilon(y_1^2(t_k) - 1)y_2(t_{k-1}))h^{q_2} - \sum_{j=1}^{k} c_j^{(q_2)} y_2(t_{k-j}). \end{aligned} \tag{5.16}$$

5.5 Problem Formulation

In this approach, the identification process is considered as a multidimensional optimization problem. In the optimization process, the parameters of a new fractional-order chaotic system FOC_E are determined by using the LS method from the operation of the original fractional-order chaotic system FOC_O. The idea is that FOC_E presents the best possible parametric affinity with FOC_O. Under such circumstances, the original fractional-order chaotic system FOC_O can be defined as follows:

$$ {}_aD_t^{\mathbf{q}}Y = F(\mathbf{Y}, \mathbf{Y_0}, \boldsymbol{\theta}), \tag{5.17}$$

where $\mathbf{Y} = [y_1, y_2, \ldots, y_m]^{\mathrm{T}}$ denotes the state vector of the system, $\mathbf{Y_0}$ symbolizes the initial state vector, $\boldsymbol{\theta} = [\theta_1, \theta_2, \ldots, \theta_m]^{\mathrm{T}}$ represents the original systematic parameter set, $\mathbf{q} = [q_1, q_2, \ldots, q_m]^{\mathrm{T}}$ for $0 < q_i < 1$ $(i \in [1, \ldots, m])$ corresponds to the fractional derivative orders and F is a generic non-linear function. On the other hand, the estimated fractional-order chaotic system FOC_E can be modeled as follows:

$$ {}_aD_t^{\hat{\mathbf{q}}}\hat{Y} = F(\hat{\mathbf{Y}}, \mathbf{Y}_0, \hat{\boldsymbol{\theta}}), \tag{5.18}$$

where $\hat{\mathbf{Y}}$, $\hat{\boldsymbol{\theta}}$ and $\hat{\mathbf{q}}$ denotes the estimated state system, the estimated systematic parameter vector and the estimated fractional orders, respectively.

Since the goal is that FOC_E presents the best possible parametric affinity with FOC_O, the problem can be approached as an optimization problem described by the following formulation:

$$ \bar{\theta}, \bar{q} = \arg \min_{(\hat{\mathbf{Y}}, \hat{\mathbf{q}}) \in \Omega} (J(\theta, q)), \tag{5.19}$$

where $\bar{\theta}, \bar{q}$ denotes the best possible parametric values obtained by the optimization process, Ω symbolizes the search space admitted for parameters ($\hat{\mathbf{Y}}$ and $\hat{\mathbf{q}}$) whereas J represents the objective function that evaluates the parametric affinity between FOC_O and FOC_E. This affinity can be computed as follows:

$$ J(\theta, q) = \frac{1}{M} \sum_{k=1}^{M} (\mathbf{Y}(k) - \hat{\mathbf{Y}}(k))^2, \tag{5.20}$$

where $\mathbf{Y}(k)$ and $\tilde{\mathbf{Y}}(k)$ represent the state values produced by the original and estimated systems, respectively. On the other hand, k denotes the sampling time point and M represents the length of data used for parameter estimation. According to the optimization problem formulated in Eq. (5.19), the parameter identification can be achieved by searching suitable values of $\hat{\mathbf{Y}}$ and $\hat{\mathbf{q}}$ within the searching space

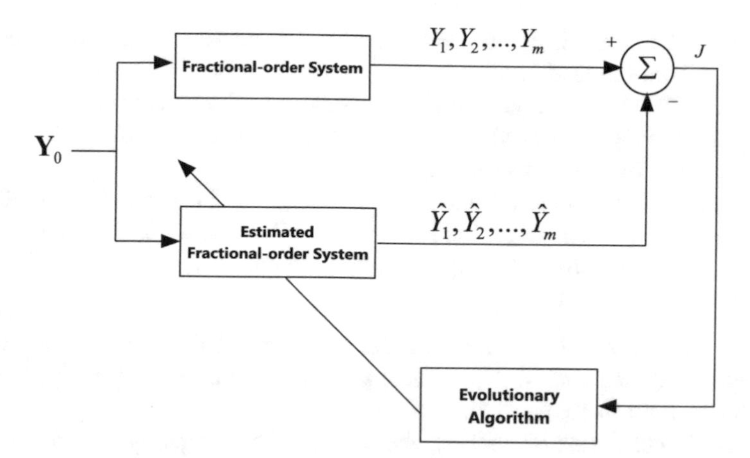

Fig. 5.2 Evolutionary algorithm for fractional-order system parameter estimation

Ω, such that the objective function has been minimized. Figure 5.2 shows the graphic representation of the identification process. Since the fractional-order Van der Pol oscillator has been chosen to test the performance of the approach, the fractional-order system maintain two different fractional derivative orders $\mathbf{q} = [q_1, q_2]^{\mathrm{T}}$ $(m = 2)$ and one systematic parameter ε.

5.6 Experimental Results

To verify the effectiveness and robustness of the approach, the fractional-order Van der Pol oscillator is chosen to test its performance. The simulations has been conducted by using MATLAB (Version 7.1, MathWorks, Natick, MA, USA) on an Intel(R) Core(TM) i7-3470 CPU, 3.2 GHz with 4 GB of RAM. In order to calculate the objective function, the number of samples is set as 300 and the step size is 0.01.

In this section, the results of the LS algorithm have been compared to those produced by the Genetic Algorithms (GA) [17], Particle Swarm Optimization (PSO) method [20], the Differential Evolution (DE) [19], and the LS method. In all comparisons, the population has been set to 40 ($N = 40$) individuals. The maximum iteration number for all functions has been set to 100. Such stop criterion has been selected to maintain compatibility to similar works reported in the literature [16].

The parameter setting for each of the algorithms in the comparison is described as follows:

1. GA: The population size has been set to 70, the crossover probability with 0.55, the mutation probability with 0.10 and number of elite individuals with 2. The roulette wheel selection and the 1-point crossover are applied.

1. PSO: In the method, $c_1 = c_2 = 2$ whereas the inertia factor (ω) is decreased linearly from 0.9 to 0.2.
2. DE: The DE/Rand/1 scheme has been employed. The parameter settings follow the instructions suggested in [31]. The crossover probability is $CR = 0.9$ whereas the weighting factor is $F = 0.8$.
3. In LS, F and L are set to 0.6 and L, respectively. Similarly, g is fixed to 20 ($N/2$), $h = 2$, $\beta = 0.6$ whereas gen and N are set to 1000 and 40, respectively. Once such parameters have been experimentally determined, they are considered for all experiments in this section.

In the experiments, the fractional-order Van der Pol Oscillator to be estimated has been configured such that $q_1 = 1.2$, $q_2 = 0.8$ and $\varepsilon = 1$. Similarly, the initial state has been set to [0.02,−0.2].

The statistical results of the best, the mean and the worst estimated parameters with the corresponding relative error values over 100 independent runs are shown in Table 5.1. From Table 5.1, it can be easily seen that the estimated values generated by the LS algorithm are closer to the actual parameter values, which means that it is more accurate than the standard GA, PSO and DE algorithms. Likewise, it can also be clearly found that the relative error values obtained by the LS algorithm are all smaller than those of the standard GA, PSO and DE algorithms, which can also

Table 5.1 Simulation result of the algorithms GA, PSO, DE and LS

	Parameter	GA	PSO	DE	LS
BEST	ε	0.9021	0.9152	0.9632	0.9978
	$\frac{\lvert\varepsilon-1\rvert}{1}$	0.0979	0.0848	0.0368	0.0022
	q_1	1.3001	1.2810	1.2210	1.2005
	$\frac{\lvert q_1-1.2\rvert}{1.2}$	0.0834	0.0675	0.0175	0.0004
	q_2	0.8702	0.8871	0.8229	0.8011
	$\frac{\lvert q_2-0.8\rvert}{0.8}$	0.0877	0.1088	0.0286	0.0013
WORST	ε	0.1731	0.1176	0.3732	0.7198
	$\frac{\lvert\varepsilon-1\rvert}{1}$	0.8269	0.8824	0.6268	0.2802
	q_1	2.1065	0.3643	1.8532	1.3075
	$\frac{\lvert q_1-1.2\rvert}{1.2}$	0.7554	0.6964	0.5443	0.0895
	q_2	0.1219	1.7643	1.2154	0.9101
	$\frac{\lvert q_2-0.8\rvert}{0.8}$	0.8476	1.2053	0.5192	0.1376
MEAN	ε	1.2131	1.2052	1.1701	1.0186
	$\frac{\lvert\varepsilon-1\rvert}{1}$	0.2131	0.2052	0.1701	0.0186
	q_1	0.9032	1.0974	1.3421	1.2654
	$\frac{\lvert q_1-1.2\rvert}{1.2}$	0.2473	0.0855	0.1186	0.0545
	q_2	0.9052	0.7229	0.7832	0.8089
	$\frac{\lvert q_2-0.8\rvert}{0.8}$	0.1315	0.0963	0.0210	0.0111

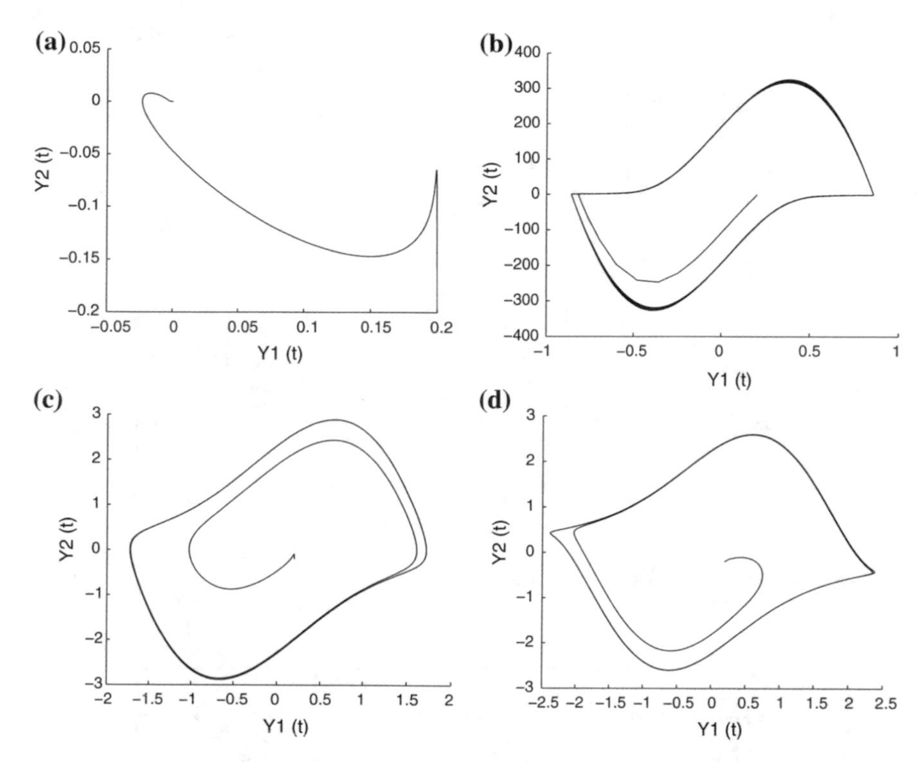

Fig. 5.3 Phase diagrams of the Van der Pol oscillator by using the mean estimated parameters for **a** GA, **b** PSO, **c** DE and **d** the LS approach

prove that the LS algorithm has a higher performance in terms of accuracy. Therefore, the estimated parameters can be closer to the true values than the GA, PSO and DE algorithms. With this evidence, it can be concluded that the LS algorithm can more efficiently identify a fractional-order systems than the other algorithms used in the comparisons. In order to show the proficiency, of the approach, Fig. 5.3 presents the phase diagrams of the Van der Pol Oscillator by using the mean estimated parameters for each method.

The convergence curves of the parameters and fitness values estimated by the set of algorithms are shown in Figs. 5.4, 5.5 and 5.6 in a single execution. From Figs. 5.4, 5.5 and 5.6, it can be clearly observed that convergence processes of the parameters and fitness values of LS algorithm are better than other algorithms. Additionally, the estimated parameter values obtained by the LS algorithm fall faster than the other algorithms.

Furthermore, Table 5.2 shows the average best solution obtained by each algorithm. The average best solution (ABS) expresses the average value of the best function evaluations that have been obtained from 100 independent executions. A non-parametric statistical significance test known as the Wilcoxon's rank sum test for independent samples [32, 33] has been conducted with an 5% significance

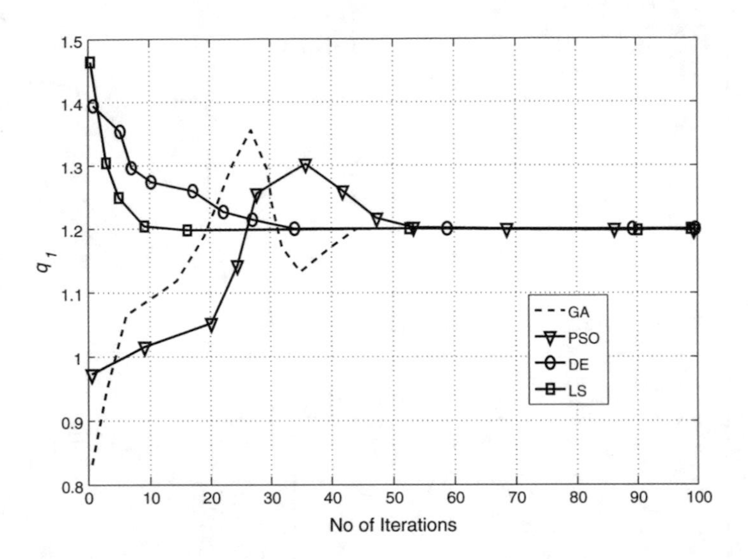

Fig. 5.4 Estimated parameter q_1 (fractional order)

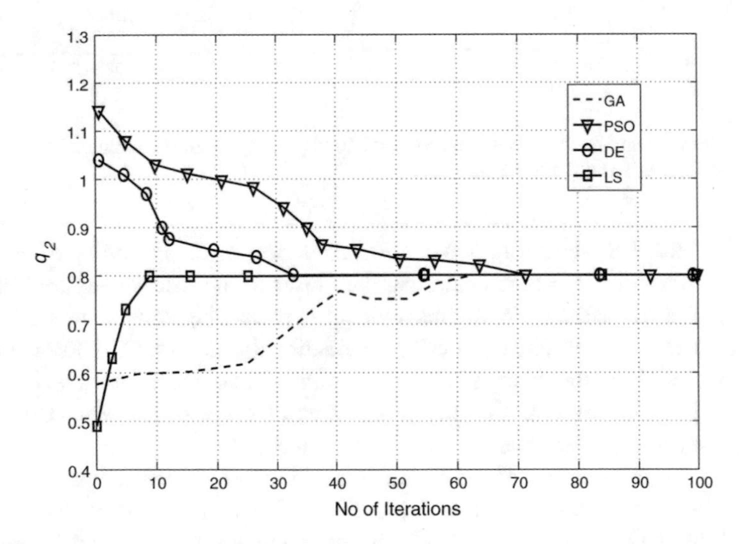

Fig. 5.5 Estimated parameter q_2 (fractional order)

level, over the "average best-solution" data of Table 5.2. Table 5.3 reports the *p*-values produced by Wilcoxon's test for the pair-wise comparison of the "average best-solution" of two groups. Such groups are formed by LS versus GA, LS versus PSO and LS versus DE. As a null hypothesis, it is assumed that there is no significant difference between mean values of the two algorithms. The alternative hypothesis considers a significant difference between the "average best-solution"

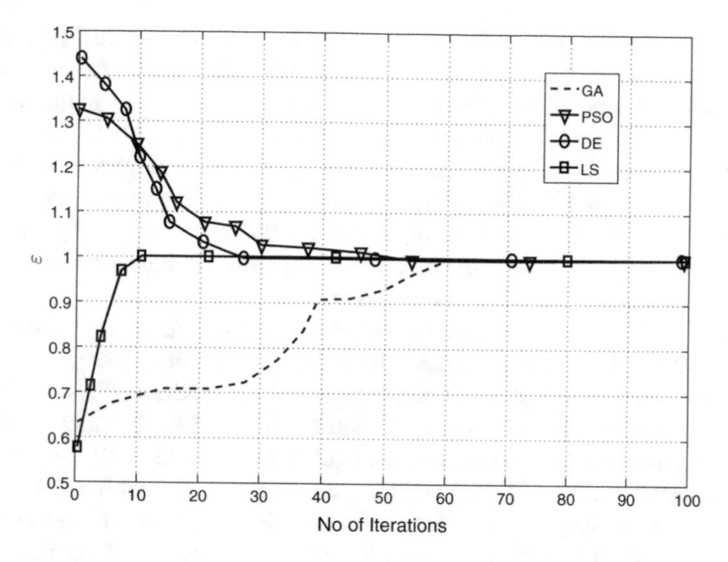

Fig. 5.6 Estimated systematic parameter ε

Table 5.2 Average best solution obtained by each algorithm GA, PSO, DE and LS

GA	PSO	DE	LS
0.2251	0.2016	0.0982	0.0126

Table 5.3 p-values produced by Wilcoxon's test that compares LS versus GA, LS versus PSO and DE over the "average best-solution" values from Table 5.3

	p-values
LS versus GA	0.00021
LS versus PSO	0.00098
LS versus DE	0.00123

values of both approaches. All p-values reported in the table are less than 0.05 (5% significance level) which is a strong evidence against the null hypothesis, indicating that the LS results are statistically significant and that it has not occurred by coincidence (i.e. due to the normal noise contained in the process).

5.7 Conclusions

Due to its multiple applications, parameter identification for fractional-order chaotic systems has attracted the interests of several research communities. In the identification, the parameter estimation process is transformed into a multidimensional optimization problem where fractional orders, as well as functional parameters of the chaotic system are considered the decision variables. Under this approach, the complexity of fractional-order chaotic systems tends to produce multimodal error surfaces for which their cost functions are significantly difficult to minimize. Several algorithms based on evolutionary computation principles have been successfully applied to identify the parameters of fractional-order chaotic systems. However, most of them maintain an important limitation, they frequently obtain sub-optimal results as a consequence of an inappropriate balance between exploration and exploitation in their search strategies.

In this chapter, an algorithm for parameter identification of fractional-order chaotic systems has been presented. In order to determine the parameters, the method uses a novel evolutionary method called Locust Search (LS) [R1] which is based on the behavior of swarms of locusts. In the LS algorithm, individuals emulate a group of locusts which interact to each other based on the biological laws of the cooperative swarm. The algorithm considers two different behaviors: solitary and social. Depending on the behavior, each individual is conducted by a set of evolutionary operators which mimics different cooperative conducts that are typically found in the swarm. Different to most of existent evolutionary algorithms, the behavioral model in the LS approach explicitly avoids the concentration of individuals in the current best positions. Such fact allows to avoid critical flaws such as the premature convergence to sub-optimal solutions and the incorrect exploration–exploitation balance.

In order to test the proficiency and robustness of the presented method, it has been compared to other algorithms based on evolutionary principles such as GA, PSO and DE. The comparison examines the identification of the fractional Van der Pol Oscillator. The results show a high performance of the proposed estimator in terms of precision and robustness.

References

1. Das, S.: Observation of Fractional Calculus in Physical System Description, pp. 101–156. Springer, New York (2011)
2. Arena, P., Caponetto, R., Fortuna, L., Porto, D.: Nonlinear Noninteger Order Circuits and Systems—An Introduction. World Scientific, Singapore (2000)
3. Rivero, M., Rogosin, S.V., Tenreiro Machado, J.A., Trujillo, J.J.: Stability of fractional order systems. Math. Probl. Eng. (2013). https://doi.org/10.1155/2013/356215
4. Diethelm, K.: An efficient parallel algorithm for the numerical solution of fractional differential equations. Fract. Calc. Appl. Anal. **14**(3), 475–490 (2011)
5. Diethelm, K., Ford, N.J.: Analysis of fractional differential equations. J. Math. Anal. Appl. **265**(2), 229–248 (2002)
6. Kilbas, A.A.A., Srivastava, H.M., Trujillo, J.J.: Theory and Applications of Fractional Differential Equations. Elsevier Science, The Netherlands (2006)
7. Podlubny, I.: Fractional Differential Equations. Academic Press, USA (1998)
8. Miller, K.S., Ross, B.: An Introduction to the Fractional Calculus and Fractional Differential Equations. Wiley, New York (1993)

9. Hu, W., Yu, Y., Zhang, S.: A hybrid artificial bee colony algorithm for parameter identification of uncertain fractional-order chaotic systems. Nonlinear Dyn. https://doi.org/10.1007/s11071-015-2251-6
10. Yu, Y., Li, H.-X., Wang, S., Yu, J.: Dynamic analysis of a fractional-order Lorenz chaotic system. Chaos, Solitons Fractals **42**, 1181–1189 (2009)
11. Petras, I.: Fractional-Order Nonlinear Systems, ISBN 978-3-642-18100-9, Springer-Verlag BerlinHeidelberg (2011)
12. Poinot, T., Trigeassou, J.-C.: Identification of fractional systems using an output error technique. Nonlinear Dyn. **38**, 133–154 (2004)
13. Nazarian, P., Haeri, M., Tavazoei, M.S.: Identifiability of fractional order systems using input output frequency contents. ISA Trans. **49**, 207–214 (2010)
14. Saha, R.S.: On Haar wavelet operational matrix of general order and its application for the numerical solution of fractional Bagley Torvik equation. Appl. Math. Comput. **218**, 5239–5248 (2012)
15. Kerschen, G., Worden, K., Vakakis, A.F., Golinval, J.C.: Past, present and future of nonlinear system identification in structural dynamics. Mech. Syst. Signal Process. **20**(3), 505–592 (2006)
16. Quaranta, G., Monti, G., Marano, G.C.: Parameters identification of Van der Pol–Duffing oscillators via particle swarm optimization and differential evolution. Mech. Syst. Signal Process. **24**, 2076–2095 (2010)
17. Zhou, S., Cao, J., Chen, Y.: Genetic algorithm-based identification of fractional-order systems. Entropy **15**, 1624–1642 (2013)
18. Hu, W., Yu, Y., Wang, S.: Parameters estimation of uncertain fractional-order chaotic systems via a modified artificial bee colony algorithm. Entropy **17**, 692–709 (2015). https://doi.org/10.3390/e17020692
19. Gao, F., Lee, X., Fei, F., Tong, H., Deng, Y., Zhao, H.: Identification time-delayed fractional order chaos with functional extrema model via differential evolution. Expert Syst. Appl. **41** (4), 1601–1608 (2014)
20. Wu, D., Ma, Z., Li, A., Zhu, Q.: Identification for fractional order rational models based on particle swarm optimization. Int. J. Comput. Appl. Technol. **41**(1/2), 53–59 (2011)
21. Tan, K.C., Chiam, S.C., Mamun, A.A., Goh, C.K.: Balancing exploration and exploitation with adaptive variation for evolutionary multi-objective optimization. Eur. J. Oper. Res. **197**, 701–713 (2009)
22. Chen, G., Low, C.P., Yang, Z.: Preserving and exploiting genetic diversity in evolutionary programming algorithms. IEEE Trans. Evol. Comput. **13**(3), 661–673 (2009)
23. Cuevas, E., González, A., Fausto, F., Zaldívar, D., Pérez-Cisneros, M.: Multithreshold segmentation by using an algorithm based on the behavior of locust swarms. Math. Probl. Eng. **2015**, Article ID 805357, 25 pages (2015). https://doi.org/10.1155/2015/805357
24. Cuevas, E., Zaldivar, D., Perez, M.: Automatic segmentation by using an algorithm based on the behavior of locust swarms. In: Applications of Evolutionary Computation in Image Processing and Pattern Recognition, Volume 100 of the series Intelligent Systems Reference Library, pp. 229–269 (2016)
25. Cuevas, E., González, A., Zaldívar, D., Pérez-Cisneros, M.: An optimisation algorithm based on the behaviour of locust swarms. Int. J. Bio-Inspired Comput. **7**(6), 402–407 (2015)
26. Miller, K.S., Ross, B.: An Introduction to the Fractional Calculus and Fractional Differential Equations. Wiley, USA (1993)
27. Dorcak, L.: Numerical models for the simulation of the fractional-order control systems (1994)
28. Quaranta, G., Monti, G., Marano, G.C.: Parameters identification of Van der Pol–Duffing oscillators via particle swarm optimization and differential evolution. Mech. Syst. Signal Process. **24**(7), 2076–2095 (2010)
29. Barbosa, R.S., Machado, J.A.T., Vinagre, B.M., Calderon, A.J.: Analysis of the Van der Pol oscillator containing derivatives of fractional order. J. Vib. Control **13** (9–10), 1291–1301 (2007)

30. Cartwright, J., Eguiluz, V., Hernandez-Garcia, E., Piro, O.: Dynamics of elastic excitable media. Int. J. Bifurcat. Chaos **9**(11), 2197–2202 (1999)
31. Cuevas, E., Zaldivar, D., Pérez-Cisneros, M., Ramírez-Ortegón, M.: Circle detection using discrete differential evolution optimization. Pattern Anal. Appl. **14**(1), 93–107 (2011)
32. Wilcoxon, F.: Individual comparisons by ranking methods. Biometrics **1**, 80–83 (1945)
33. Garcia, S., Molina, D., Lozano, M., Herrera, F.: A study on the use of non-parametric tests for analyzing the evolutionary algorithms' behaviour: a case study on the CEC '2005, Special session on real parameter optimization. J. Heurist (2008). https://doi.org/10.1007/s10732-008-9080-4

Chapter 6
The States of Matter Search (SMS)

The capacity of a metaheuristic method to attain the global optimal solution maintains an explicit dependency on its potential to find a good balance between exploitation and exploration of the search strategy. Several works have been developed with many interesting metaheuristic methods which, at original versions, introduce operations without considering the exploration–exploitation ratio. In this chapter, the States of Matter Search (SMS) is analyzed. The SMS method is based on the emulation of the states of matter concepts. In SMS, candidate solutions represent molecules which interact among them by employing metaheuristic operators under the physical laws of the thermal-energy motion phenomena. The method is designed assuming that each state of matter defines a different exploration–exploitation ratio. The optimization process is subdivided into three stages which follow the different states of matter: gas, liquid and solid. In each phase, molecules (individuals) present a different kind of movement. Starting from the gas phase (only exploration), the method adjusts the rates of exploration and exploitation until the solid state (only exploitation) is attained. As a result, the algorithm can considerably enhance the equilibrium between exploration–exploitation, yet preserving the excellent search aptitudes of a metaheuristic approach.

6.1 Introduction

Global optimization [1] has delivered applications for many areas of science, engineering, economics and others, where mathematical modelling is used [2]. In general, the goal is to find a global optimum for an objective function which is defined over a given search space. Global optimization algorithms are usually broadly divided into deterministic and stochastic [3]. Since deterministic methods only provide a theoretical guarantee of locating a local minimum of the objective function, they often face great difficulties in solving global optimization problems [4]. On the other hand, evolutionary algorithms are usually faster in locating a

E. Cuevas et al., *Advances in Metaheuristics Algorithms: Methods and Applications*, Studies in Computational Intelligence 775,
https://doi.org/10.1007/978-3-319-89309-9_6

global optimum [5]. Moreover, stochastic methods adapt easily to black-box formulations and extremely ill-behaved functions, whereas deterministic methods usually rest on at least some theoretical assumptions about the problem formulation and its analytical properties (such as Lipschitz continuity) [6].

Evolutionary algorithms, which are considered as members of the stochastic group, have been developed by a combination of rules and randomness that mimics several natural phenomena. Such phenomena include evolutionary processes such as the Evolutionary Algorithm (EA) proposed by Fogel et al. [7], De Jong [8], and Koza [9], the Genetic Algorithm (GA) proposed by Holland [10] and Goldberg [11], the Artificial Immune System proposed by De Castro and Von Zuben [12] and the Differential Evolution Algorithm (DE) proposed by Price and Storn [13]. Some other methods which are based on physical processes include the Simulated Annealing proposed by Kirkpatrick et al. [14], the Electromagnetism-like Algorithm proposed by İlker et al. [15] and the Gravitational Search Algorithm proposed by Rashedia et al. [16]. Also, there are other methods based on the animal-behavior phenomena such as the Particle Swarm Optimization (PSO) algorithm proposed by Kennedy and Eberhart [17] and the Ant Colony Optimization (ACO) algorithm proposed by Dorigo et al. [18].

Every EA needs to address the issue of exploration–exploitation of the search space. Exploration is the process of visiting entirely new points of a search space whilst exploitation is the process of refining those points within the neighborhood of previously visited locations, in order to improve their solution quality. Pure exploration degrades the precision of the evolutionary process but increases its capacity to find new potential solutions. On the other hand, pure exploitation allows refining existent solutions but adversely driving the process to local optimal solutions. Therefore, the ability of an EA to find a global optimal solution depends on its capacity to find a good balance between the exploitation of found-so-far elements and the exploration of the search space [19]. So far, the exploration–exploitation dilemma has been an unsolved issue within the framework of EA.

Although PSO, DE and GSA are considered the most popular algorithms for many optimization applications, they fail in finding a balance between exploration and exploitation [20]; in multimodal functions, they do not explore the whole region effectively and often suffers premature convergence or loss of diversity. In order to deal with this problem, several proposals have been suggested in the literature [21–46]. In most of the approaches, exploration and exploitation is modified by the proper settings of control parameters that have an influence on the algorithm's search capabilities [47]. One common strategy is that EAs should start with exploration and then gradually change into exploitation [48]. Such a policy can be easily described with deterministic approaches where the operator that controls the individual diversity decreases along with the evolution. This is generally correct, but such a policy tends to face difficulties when solving certain problems with multimodal functions that hold many optima, since a premature takeover of exploitation over exploration occurs. Some approaches that use this strategy can be found in [21–29]. Other works [30–34] use the population size as reference to change the balance between exploration and exploitation. A larger population size implies a wider exploration while a smaller population demands a shorter search. Although this technique delivers an easier way to keep diversity, it often represents an unsatisfactory solution. An improper handling of large populations might converge to only one point, despite introducing more function evaluations. Recently, new operators have been added to several traditional evolutionary algorithms in order to improve their original exploration–exploitation capability. Such operators diversify particles whenever they concentrate on a local optimum. Some methods that employ this technique are discussed in [35–46].

Either of these approaches is necessary but not sufficient to tackle the problem of the exploration–exploitation balance. Modifying the control parameters during the evolution process without the incorporation of new operators to improve the population diversity, makes the algorithm defenseless against the premature convergence and may result in poor exploratory characteristics of the algorithm [48]. On the other hand, incorporating new operators without modifying the control parameters leads to increase the computational cost, weakening the exploitation process of candidate regions [39]. Therefore, it does seem reasonable to incorporate both of these approaches into a single algorithm.

In this chapter, a novel nature-inspired algorithm, known as the States of Matter Search (SMS) is analyzed for solving global optimization problems. The SMS algorithm is based on the simulation of the states of matter phenomenon. In SMS, individuals emulate molecules which interact to each other by using evolutionary operations based on the physical principles of the thermal-energy motion mechanism. Such operations allow the increase of the population diversity and avoid the concentration of particles within a local minimum. The SMS approach combines the use of the defined operators with a control strategy that modifies the parameter setting of each operation during the evolution process. In contrast to other approaches that enhance traditional EA algorithms by incorporating some procedures for balancing the exploration–exploitation rate, the SMS algorithm naturally delivers such property as a result of mimicking the states of matter phenomenon. The algorithm is devised by considering each state of matter at one different exploration–exploitation ratio. Thus, the evolutionary process is divided into three stages which emulate the three states of matter: gas, liquid and solid. At each state, molecules (individuals) exhibit different behaviors. Beginning from the gas state (pure exploration), the algorithm modifies the intensities of exploration and exploitation until the solid state (pure exploitation) is reached. As a result, the approach can substantially improve the balance between exploration–exploitation, yet preserving the good search capabilities of an evolutionary approach. To illustrate the proficiency and robustness of the SMS algorithm, it has been compared to other well-known evolutionary methods including recent variants that incorporate diversity preservation schemes. The comparison examines several standard benchmark functions which are usually employed within the EA field. Experimental results show that the SMS method achieves good performance over its counterparts as a consequence of its better exploration–exploitation capability.

This chapter is organized as follows. Section 6.2 introduces basic characteristics of the three states of matter. In Sect. 6.3, the novel SMS algorithm and its characteristics are both described. Section 6.4 presents experimental results and a comparative study. Finally, in Sect. 6.5, some conclusions are discussed.

6.2 States of Matter

The matter can take different phases which are commonly known as states. Traditionally, three states of matter are known: solid, liquid, and gas. The differences among such states are based on forces which are exerted among particles composing a material [49].

In the gas phase, molecules present enough kinetic energy so that the effect of intermolecular forces is small (or zero for an ideal gas), while the typical distance between neighboring molecules is greater than the molecular size. A gas has no definite shape or volume, but occupies the entire container in which it is confined. Figure 6.1a shows the movements exerted by particles in a gas state. The movement experimented by the molecules represent the maximum permissible displacement ρ_1 among particles [50]. In a liquid state, intermolecular forces are more restrictive than those in the gas state. The molecules have enough energy to move relatively to each other still keeping a mobile structure. Therefore, the shape of a liquid is not definite but is determined by its container. Figure 6.1b presents a particle movement ρ_2 within a liquid state. Such movement is smaller than those considered by the gas state but larger than the solid state [51]. In the solid state, particles (or molecules) are packed together closely with forces among particles being strong enough so that the particles cannot move freely but only vibrate. As a result, a solid has a stable, definite shape and a definite volume. Solids can only change their shape by force, as when they are broken or cut. Figure 6.1c shows a molecule configuration in a solid state. Under such conditions, particles are able to vibrate (being perturbed) considering a minimal ρ_3 distance [50].

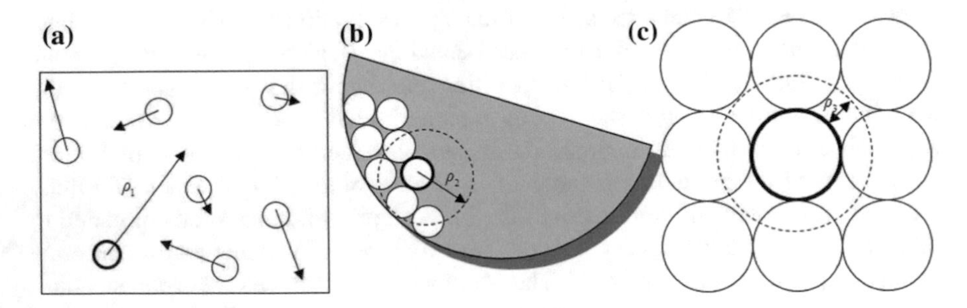

Fig. 6.1 Different states of matter: **a** gas, **b** liquid, and **c** solid

In this chapter, the States of Matter Search (SMS) is analyzed for solving global optimization problems. The SMS algorithm is based on the simulation of the states of matter phenomenon that considers individuals as molecules which interact to each other by using evolutionary operations based on the physical principles of the thermal-energy motion mechanism. The algorithm is devised by considering each state of matter at one different exploration–exploitation ratio. Thus, the evolutionary process is divided into three stages which emulate the three states of matter: gas, liquid and solid. In each state, individuals exhibit different behaviors.

6.3 Fitness Approximation Method

6.3.1 Definition of Operators

In the approach, individuals are considered as molecules whose positions on a multidimensional space are modified as the algorithm evolves. The movement of such molecules is motivated by the analogy to the motion of thermal-energy.

The velocity and direction of each molecule's movement are determined by considering the collision, the attraction forces and the random phenomena experimented by the molecule set [52]. In our approach, such behaviors have been implemented by defining several operators such as the direction vector, the collision and the random positions operators, all of which emulate the behavior of actual physics laws.

The direction vector operator assigns a direction to each molecule in order to lead the particle movement as the evolution process takes place. On the other side, the collision operator mimics those collisions that are experimented by molecules as they interact to each other. A collision is considered when the distance between two molecules is shorter than a determined proximity distance. The collision operator is thus implemented by interchanging directions of the involved molecules. In order to simulate the random behavior of molecules, the SMS algorithm generates random positions following a probabilistic criterion that considers random locations within a feasible search space.

The next section presents all operators that are used in the algorithm. Although such operators are the same for all the states of matter, they are employed over a different configuration set depending on the particular state under consideration.

Direction vector

The direction vector operator mimics the way in which molecules change their positions as the evolution process develops. For each n-dimensional molecule $\mathbf{p}_i$ from the population $\mathbf{P}$, it is assigned an n-dimensional direction vector $\mathbf{d}_i$ which stores the vector that controls the particle movement. Initially, all the direction vectors $\left(\mathbf{D} = \{\mathbf{d}_1, \mathbf{d}_2, \ldots, \mathbf{d}_{N_p}\}\right)$ are randomly chosen within the range of [−1,1].

As the system evolves, molecules experiment several attraction forces. In order to simulate such forces, the SMS algorithm implements the attraction phenomenon by moving each molecule towards the best so-far particle. Therefore, the new direction vector for each molecule is iteratively computed considering the following model:

$$\mathbf{d}_i^{k+1} = \mathbf{d}_i^k \cdot \left(1 - \frac{k}{gen}\right) \cdot 0.5 + \mathbf{a}_i, \tag{6.1}$$

where $\mathbf{a}_i$ represents the attraction unitary vector calculated as $\mathbf{a}_i = (\mathbf{p}^{best} - \mathbf{p}_i)/\|\mathbf{p}^{best} - \mathbf{p}_i\|$, being $\mathbf{p}^{best}$ the best individual seen so-far, while $\mathbf{p}_i$ is the molecule i of population $\mathbf{P}$. k represents the iteration number whereas gen involves the total iteration number that constitutes the complete evolution process.

Under this operation, each particle is moved towards a new direction which combines the past direction, which was initially computed, with the attraction vector over the best individual seen so-far. It is important to point out that the relative importance of the past direction decreases as the evolving process advances. This particular type of interaction avoids the quick concentration of information among particles and encourages each particle to search around a local candidate region in its neighborhood, rather than interacting to a particle lying at distant region of the domain. The use of this scheme has two advantages: first, it prevents the particles from moving toward the global best position in early stages of algorithm and thus makes the algorithm less susceptible to premature convergence; second, it encourages particles to explore their own neighborhood thoroughly, just before they converge towards a global best position. Therefore, it provides the algorithm with local search ability enhancing the exploitative behavior.

In order to calculate the new molecule position, it is necessary to compute the velocity $\mathbf{v}_i$ of each molecule by using:

$$\mathbf{v}_i = \mathbf{d}_i \cdot v_{init} \tag{6.2}$$

being v_{init} the initial velocity magnitude which is calculated as follows:

$$v_{init} = \frac{\sum_{j=1}^{n} (b_j^{high} - b_j^{low})}{n} \cdot \beta \tag{6.3}$$

where b_j^{low} and b_j^{high} are the low j parameter bound and the upper j parameter bound respectively, whereas $\beta \in [0, 1]$.

Then, the new position for each molecule is updated by:

$$p_{i,j}^{k+1} = p_{i,j}^{k} + v_{i,j} \cdot \text{rand}(0,1) \cdot \rho \cdot (b_j^{high} - b_j^{low}) \tag{6.4}$$

where $0.5 \leq \rho \leq 1$.

Collision

The collision operator mimics the collisions experimented by molecules while they interact to each other. Collisions are calculated if the distance between two molecules is shorter than a determined proximity value. Therefore, if $\|\mathbf{p}_i - \mathbf{p}_q\| < r$, a collision between molecules i and q is assumed; otherwise, there is no collision, considering $i, q \in \{1, \ldots, N_p\}$ such that $i \neq q$. If a collision occurs, the direction vector for each particle is modified by interchanging their respective direction vectors as follows:

$$\mathbf{d}_i = \mathbf{d}_q \text{ and } \mathbf{d}_q = \mathbf{d}_i \tag{6.5}$$

The collision radius is calculated by:

$$r = \frac{\sum_{j=1}^{n} (b_j^{high} - b_j^{low})}{n} \cdot \alpha \tag{6.6}$$

where $\alpha \in [0,1]$.

Under this operator, a spatial region enclosed within the radius r is assigned to each particle. In case the particle regions collide to each other, the collision operator acts upon particles by forcing them out of the region. The radio r and the collision operator provide the ability to control diversity throughout the search process. In other words, the rate of increase or decrease of diversity is predetermined for each stage. Unlike other diversity-guided algorithms, it is not necessary to inject diversity into the population when particles gather around a local optimum because the diversity will be preserved during the overall search process. The collision incorporation therefore enhances the exploratory behavior in the SMS approach.

Random positions

In order to simulate the random behavior of molecules, the SMS algorithm generates random positions following a probabilistic criterion within a feasible search space.

For this operation, a uniform random number r_m is generated within the range [0,1]. If r_m is smaller than a threshold H, a random molecule's position is generated; otherwise, the element remains with no change. Therefore such operation can be modeled as follows:

$$p_{i,j}^{k+1} = \begin{cases} b_j^{low} + \text{rand}(0,1) \cdot (b_j^{high} - b_j^{low}) & \text{with probability } H \\ p_{i,j}^{k+1} & \text{with probability } (1 - H) \end{cases} \tag{6.7}$$

where $i \in \{1,\ldots,N_p\}$ and $j \in \{1,\ldots,n\}$.

Best Element Updating

Despite this updating operator does not belong to State of Matter metaphor, it is used to simply store the best so-far solution. In order to update the best molecule $\mathbf{p}^{best}$ seen so-far, the best found individual from the current k population $\mathbf{p}^{best,k}$ is compared to the best individual $\mathbf{p}^{best,k-1}$ of the last generation. If $\mathbf{p}^{best,k}$ is better than $\mathbf{p}^{best,k-1}$ according to its fitness value, $\mathbf{p}^{best}$ is updated with $\mathbf{p}^{best,k}$, otherwise $\mathbf{p}^{best}$ remains with no change. Therefore, $\mathbf{p}^{best}$ stores the best historical individual found so-far.

6.3.2 SMS Algorithm

The overall SMS algorithm is composed of three stages corresponding to the three States of Matter: the gas, the liquid and the solid state. Each stage has its own behavior. In the first stage (gas state), exploration is intensified whereas in the second one (liquid state) a mild transition between exploration and exploitation is executed. Finally, in the third phase (solid state), solutions are refined by emphasizing the exploitation process.

General procedure

At each stage, the same operations are implemented. However, depending on which state is referred, they are employed considering a different parameter configuration. The general procedure in each state is shown as pseudo-code in Algorithm 1. Such procedure is composed by five steps and maps the current population $\mathbf{P}^k$ to a new population $\mathbf{P}^{k+1}$. The algorithm receives as input the current population $\mathbf{P}^k$ and the configuration parameters ρ, β, α, and H, whereas it yields the new population $\mathbf{P}^{k+1}$.

The complete algorithm

The complete algorithm is divided into four different parts. The first corresponds to the initialization stage, whereas the last three represent the States of Matter. All the optimization process, which consists of a *gen* number of iterations, is organized into three different asymmetric phases, employing 50% of all iterations for the gas state

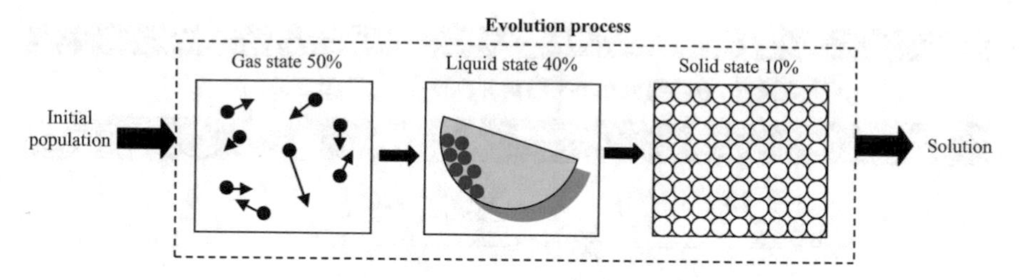

Fig. 6.2 Evolution process in the SMS approach

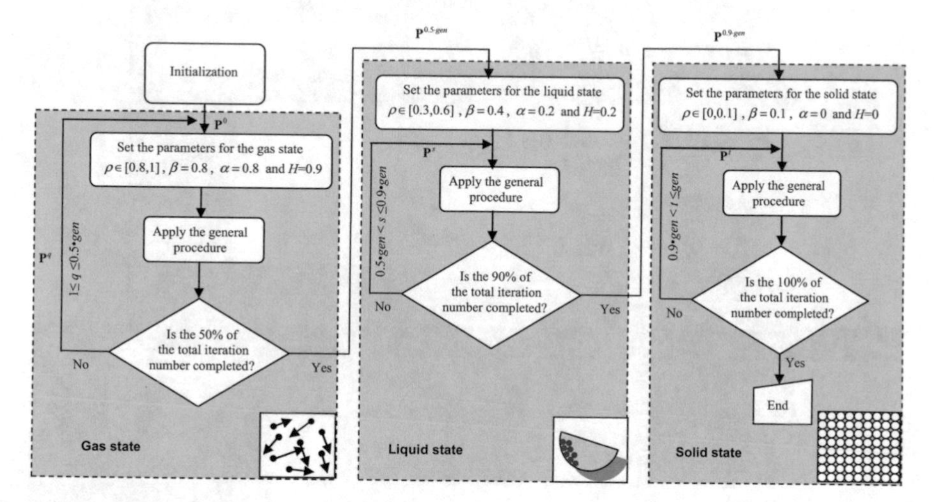

Fig. 6.3 Data flow in the complete SMS algorithm

(exploration), 40% for the liquid state (exploration–exploitation) and 10% for the solid state (exploitation). The overall process is graphically described by Fig. 6.2. At each state, the same general procedure (see Algorithm 1) is iteratively used considering the particular configuration predefined for each State of Matter. Figure 6.3 shows the data flow for the complete SMS algorithm.

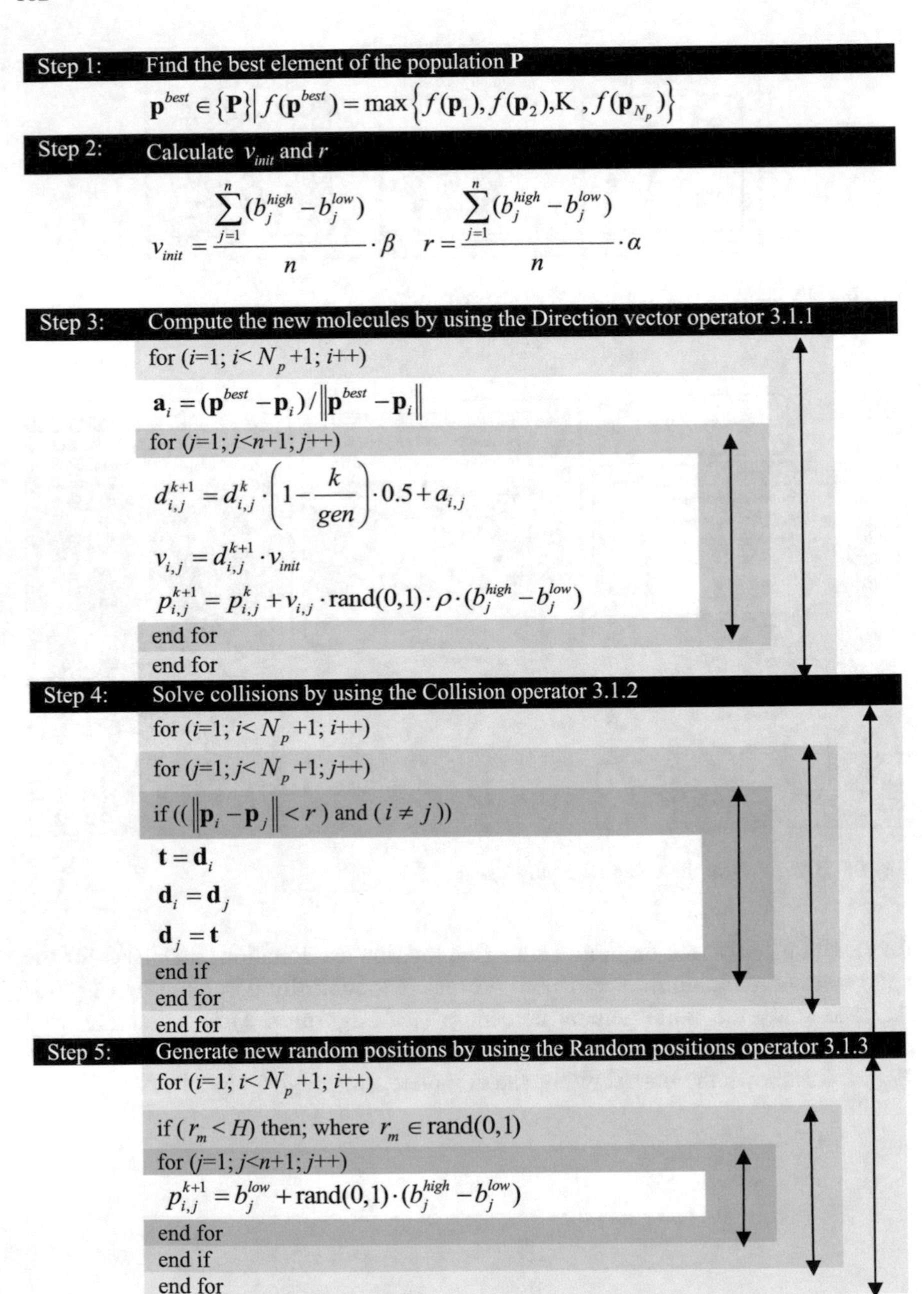

Algorithm 1. General procedure executed by all the states of matter.

Initialization

The algorithm begins by initializing a set **P** of N_p molecules ($\mathbf{P} = \{\mathbf{p}_1, \mathbf{p}_2, \ldots, \mathbf{p}_{N_p}\}$). Each molecule position $\mathbf{p}_i$ is a n-dimensional vector containing the parameter values to be optimized. Such values are randomly and uniformly distributed between the pre-specified lower initial parameter bound b_j^{low} and the upper initial parameter bound b_j^{high}, just as it is described by the following expressions:

$$p_{i,j}^{0} = b_j^{low} + \text{rand}(0,1) \cdot (b_j^{high} - b_j^{low}) \quad j = 1, 2, \ldots, n;\ i = 1, 2, \ldots, N_p, \tag{6.8}$$

where j and i, are the parameter and molecule index respectively whereas zero indicates the initial population. Hence, p_i^j is the jth parameter of the ith molecule.

Gas state

In the gas state, molecules experiment severe displacements and collisions. Such state is characterized by random movements produced by non-modeled molecule phenomena [52]. Therefore, the ρ value from the direction vector operator is set to a value near to one so that the molecules can travel longer distances. Similarly, the H value representing the random positions operator is also configured to a value around one, in order to allow the random generation for other molecule positions. The gas state is the first phase and lasts for the 50% of all iterations which compose the complete optimization process. The computational procedure for the gas state can be summarized as follows:

Step 1: Set the parameters $\rho \in [0.8,1]$, $\beta = 0.8$, $\alpha = 0.8$ and $H = 0.9$ being consistent with the gas state.
Step 2: Apply the general procedure which is illustrated in Algorithm 1.
Step 3: If the 50% of the total iteration number is completed ($1 \leq k \leq 0.5 \cdot gen$), then the process continues to the liquid state procedure; otherwise go back to step 2.

Liquid state

Although molecules currently at the liquid state exhibit restricted motion in comparison to the gas state, they still show a higher flexibility with respect to the solid state. Furthermore, the generation of random positions which are produced by non-modeled molecule phenomena is scarce [53]. For this reason, the ρ value from the direction vector operator is bounded to a value between 0.3 and 0.6. Similarly, the random position operator H is configured to a value near to cero in order to allow the random generation of fewer molecule positions. In the liquid state, collisions are also less common than in gas state, so the collision radius, that is controlled by α, is set to a smaller value in comparison to the gas state. The liquid state is the second phase and lasts the 40% of all iterations which compose the complete optimization process. The computational procedure for the liquid state can be summarized as follows:

Step 4: Set the parameters $\rho \in [0.3, 0.6]$, $\beta = 0.4$, $\alpha = 0.2$ and $H = 0.2$ being consistent with the liquid state.
Step 5: Apply the general procedure that is defined in Algorithm 1.
Step 6: If the 90% (50% from the gas state and 40% from the liquid state) of the total iteration number is completed $(0.5 \cdot gen < k \leq 0.9 \cdot gen)$, then the process continues to the solid state procedure; otherwise go back to step 5.

Solid state

In the solid state, forces among particles are stronger so that particles cannot move freely but only vibrate. As a result, effects such as collision and generation of random positions are not considered [52]. Therefore, the ρ value of the direction vector operator is set to a value near to zero indicating that the molecules can only vibrate around their original positions. The solid state is the third phase and lasts for the 10% of all iterations which compose the complete optimization process. The computational procedure for the solid state can be summarized as follows:

Step 7: Set the parameters $\rho \in [0.0, 0.1]$ and $\beta = 0.1$, $\alpha = 0$ and $H = 0$ being consistent with the solid state.
Step 8: Apply the general procedure that is defined in Algorithm 1.
Step 9: If the 100% of the total iteration number is completed $(0.9 \cdot gen < k \leq gen)$, the process is finished; otherwise go back to step 8.

It is important to clarify that the use of this particular configuration ($\alpha = 0$ and $H = 0$) disables the collision and generation of random positions operators which have been illustrated in the general procedure.

6.4 Experimental Results

A comprehensive set of 24 functions, collected from Refs. [54–61], has been used to test the performance of the SMS approach. Tables 6.8, 6.9, 6.10 and 6.11 in the Appendix present the benchmark functions used in our experimental study. Such functions are classified into four different categories: Unimodal test functions (Table 6.8), multimodal test functions (Table 6.9), multimodal test functions with fixed dimensions (Table 6.10) and functions proposed for the GECCO contest (Table 6.11). In such tables, n indicates the dimension of the function, f_{opt} the optimum value of the function and S the subset of R^n. The function optimum position ($\mathbf{x}_{opt}$) for $f_1, f_2, f_4, f_6, f_7 f_{10}, f_{11}$ and f_{14} is at $\mathbf{x}_{opt} = [0]^n$, for f_3, f_8 and f_9 is at $\mathbf{x}_{opt} = [1]^n$, for f_5 is at $\mathbf{x}_{opt} = [420.96]^n$, for f_{18} is at $\mathbf{x}_{opt} = [0]^n$, for f_{12} is at $\mathbf{x}_{opt} = [0.0003075]^n$ and for f_{13} is at $\mathbf{x}_{opt} = [-3.32]^n$. In case of functions contained in

Table 6.11, the $\mathbf{x}^{opt}$ and f_{opt} values have been set to default values which have been obtained from the Matlab© implementation for GECCO competitions, as it is provided in [59]. A detailed description of optimum locations is given in Appendix.

6.4.1 Performance Comparison to Other Meta-Heuristic Algorithms

We have applied the SMS algorithm to 24 functions whose results have been compared to those produced by the Gravitational Search Algorithm (GSA) [16], the Particle Swarm Optimization (PSO) method [17] and the Differential Evolution (DE) algorithm [13]. These are considered as the most popular algorithms in many optimization applications. In order to enhance the performance analysis, the PSO algorithm with a territorial diversity-preserving scheme (TPSO) [39] has also been added into the comparisons. TPSO is considered a recent PSO variant that incorporates a diversity preservation scheme in order to improve the balance between exploration and exploitation. In all comparisons, the population has been set to 50. The maximum iteration number for functions in Tables 6.8, 6.9 and 6.11 has been set to 1000 and for functions in Table 6.10 has been set to 500. Such stop criterion has been selected to maintain compatibility to similar works reported in the literature [4, 16].

The parameter setting for each algorithm in the comparison is described as follows:

1. GSA [16]: The parameters are set to $G_o = 100$ and $\alpha = 20$; the total number of iterations is set to 1000 for functions f_1 to f_{11} and 500 for functions f_{12} to f_{14}. The total number of individuals is set to 50. Such values are the best parameter set for this algorithm according to [16].
2. PSO [17]: The parameters are set to $c_1 = 2$ and $c_2 = 2$; besides, the weight factor decreases linearly from 0.9 to 0.2.
3. DE [13]: The DE/Rand/1 scheme is employed. The crossover probability is set to $CR = 0.9$ and the weight factor is set to $F = 0.8$.
4. TPSO [39]: The parameter α has been set to 0.5. Such value is found to be the best configuration according to [39]. The algorithm has been tuned according to the set of values which have been originally proposed by its own reference.

The experimental comparison between metaheuristic algorithms with respect to SMS has been developed according to the function-type classification as follows:

1. Unimodal test functions (Table 6.8).
2. Multimodal test functions (Table 6.9).
3. Multimodal test functions with fixed dimension (Table 6.10).
4. Test functions from the GECCO contest (Table 6.11).

Unimodal test functions

This experiment is performed over the functions presented in Table 6.8. The test compares the SMS to other algorithms such as GSA, PSO, DE and TPSO. The results for 30 runs are reported in Table 6.1 considering the following performance indexes: the Average Best-so-far (AB) solution, the Median Best-so-far (MB) and the Standard Deviation (SD) of best-so-far solution. The best outcome for each function is boldfaced. According to this table, SMS delivers better results than GSA, PSO, DE and TPSO for all functions. In particular, the test remarks the largest difference in performance which is directly related to a better trade-off between exploration and exploitation. Just as it is illustrated by Fig. 6.4, SMS, DE and GSA have similar convergence rates at finding the optimal minimal, yet faster than PSO and TPSO.

A non-parametric statistical significance proof known as the Wilcoxon's rank sum test for independent samples [62, 63] has been conducted over the "average best-so-far" (AB) data of Table 6.1, with an 5% significance level. Table 6.2 reports the p-values produced by Wilcoxon's test for the pair-wise comparison of the "average best so-far" of four groups. Such groups are formed by SMS versus GSA, SMS versus PSO, SMS versus DE and SMS versus TPSO. As a null hypothesis, it is assumed that there is no significant difference between mean values of the two algorithms. The alternative hypothesis considers a significant difference between the "average best-so-far" values of both approaches. All p-values reported in Table 6.2 are less than 0.05 (5% significance level) which is a strong evidence against the null hypothesis. Therefore, such evidence indicates that SMS results are statistically significant and that it has not occurred by coincidence (i.e. due to common noise contained in the process).

Table 6.1 Minimization result of benchmark functions of Table 6.8 with $n = 30$

		SMS	GSA	PSO	DE	TPSO
f_1	AB	**4.68457E−16**	1.3296E−05	0.873813333	0.186584241	0.100341256
	MB	4.50542E−16	7.46803E−06	4.48139E−12	0.189737658	0.101347821
	SD	1.23694E−16	1.45053E−05	4.705628811	0.039609704	0.002421043
f_2	AB	**0.033116745**	0.173618066	12.83021186	54.85755486	0.103622066
	MB	1.02069E−08	0.159932758	12.48059177	54.59915941	0.122230612
	SD	0.089017369	0.122230612	3.633980625	4.506836836	0.006498124
f_3	AB	**19.64056183**	32.83253962	33399.69716	46898.34558	21.75247912
	MB	26.87914282	27.65055745	565.0810149	43772.19502	28.45741892
	SD	11.8115879	19.11361524	43099.34439	15697.6366	14.56258711
f_4	AB	**8.882513655**	9.083435186	15.05362961	12.83391861	13.98432748
	MB	9.016816582	9.150769929	13.91301428	12.89762202	14.01237836
	SD	0.442124359	0.499181789	4.790792877	0.542197802	1.023476914

Maximum number of iterations = 1000
Bold elements respresent the best values

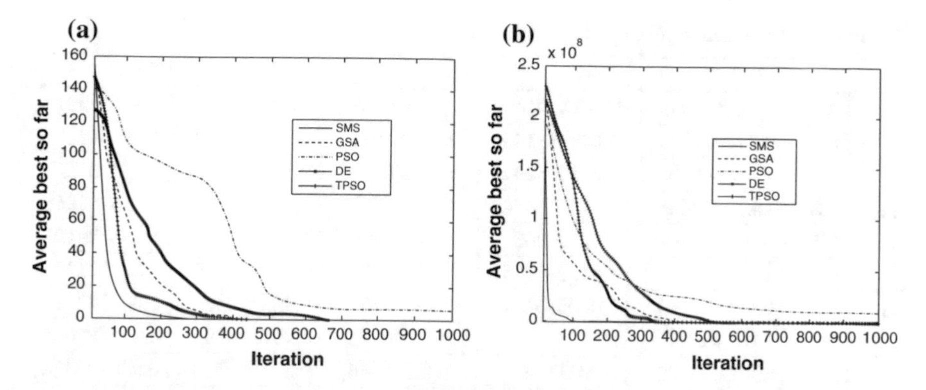

Fig. 6.4 Convergence rate comparison of GSA, PSO, DE, SMS and TPSO for minimization of **a** f_1 and **b** f_3 considering $n = 30$

Table 6.2 *p*-values produced by Wilcoxon's test comparing SMS versus PSO, SMS versus GSA, SMS versus DE and SMS versus TPSO over the "average best-so-far" (AB) values from Table 6.3

SMS versus	PSO	GSA	DE	TPSO
f_1	3.94×10^{-5}	7.39×10^{-4}	1.04×10^{-6}	4.12×10^{-4}
f_2	5.62×10^{-5}	4.92×10^{-4}	2.21×10^{-6}	3.78×10^{-4}
f_3	6.42×10^{-8}	7.11×10^{-7}	1.02×10^{-4}	1.57×10^{-4}
f_4	1.91×10^{-8}	7.39×10^{-4}	1.27×10^{-6}	4.22×10^{-4}

Multimodal test functions

Multimodal functions represent a good optimization challenge as they possess many local minima (Table 6.9). In the case of multimodal functions, final results are very important since they reflect the algorithm's ability to escape from poor local optima and to locate a near-global optimum. Experiments using f_5 to f_{11} are quite relevant as the number of local minima for such functions increases exponentially as their dimensions increase. The dimension of such functions is set to 30. The results are averaged over 30 runs, reporting the performance index for each function in Table 6.3 as follows: the Average Best-so-far (AB) solution, the Median Best-so-far (MB) and the Standard Deviation (SD) best-so-far (the best result for each function is highlighted). Likewise, *p*-values of the Wilcoxon signed-rank test of 30 independent runs are listed in Table 6.4.

In the case of functions f_8, f_9, f_{10} and f_{11}, SMS yields much better solutions than other methods. However, for functions f_5, f_6 and f_7, SMS produces similar results to GSA and TPSO. The Wilcoxon rank test results, which are presented in Table 6.4, demonstrate that SMS performed better than GSA, PSO, DE and TPSO considering four functions f_8–f_{11}, whereas, from a statistical viewpoint, there is no difference between results from SMS, GSA and TPSO for f_5, f_6 and f_7. The progress of the "average best-so-far" solution over 30 runs for functions f_5 and f_{11} is shown by Fig. 6.5.

Table 6.3 Minimization result of benchmark functions in Table 6.9 with n = 30

		SMS	GSA	PSO	DE	TPSO
f_5	AB	**1756.862345**	9750.440145	4329.650468	4963.600685	1893.673916
	MB	0.070624076	9838.388135	4233.282929	5000.245932	50.23617893
	SD	1949.048601	405.1365297	699.7276454	202.2888921	341.2367823
f_6	AB	**10.95067665**	15.18970458	130.5959941	194.6220253	18.56962853
	MB	0.007142491	13.9294268	129.4942809	196.1369499	1.234589423
	SD	14.38387472	4.508037915	27.87011038	9.659933059	7.764931264
f_7	AB	**0.000299553**	0.000575111	0.19630233	0.98547042	0.002348619
	MB	8.67349E−05	0	0.011090373	0.991214493	0.000482084
	SD	0.000623992	0.0021752	0.702516846	0.031985616	0.000196428
f_8	AB	**1.35139E−05**	2.792846799	1450.666769	304.6986718	1.753493426
	MB	7.14593E−06	2.723230534	0.675050254	51.86661185	1.002364819
	SD	2.0728E−05	1.324814757	1708.798785	554.2231579	0.856294537
f_9	AB	**0.002080591**	14.49783478	136.6888694	67251.29956	5.284029512
	MB	0.000675275	9.358377669	7.00288E−05	37143.43153	0.934751939
	SD	0.003150999	18.02351657	7360.920758	63187.52749	1.023483601
f_{10}	AB	**0.003412411**	40.59204902	365.7806149	822.087914	9.636393364
	MB	0.003164797	39.73690704	359.104488	829.1521586	0.362322274
	SD	0.001997493	11.46284891	148.9342039	81.93476435	2.194638533
f_{11}	AB	**0.199873346**	1.121397135	0.857971914	3.703467688	0.452738336
	MB	0.199873346	1.114194975	0.499967033	3.729096071	0.124948295
	SD	0.073029674	0.271747312	1.736399225	0.278860779	0.247510642

Maximum number of iterations = 1000
Bold elements represents the best values

Table 6.4 p-values produced by Wilcoxon's test comparing SMS versus GSA, SMS versus PSO, SMS versus DE and SMS versus TPSO over the "average best-so-far" (AB) values from Table 6.3

SMS versus	GSA	PSO	DE	TPSO
f_5	0.087	8.38×10^{-4}	4.61×10^{-4}	0.058
f_6	0.062	1.92×10^{-9}	9.97×10^{-8}	0.012
f_7	0.055	4.21×10^{-5}	3.34×10^{-4}	0.061
f_8	7.74×10^{-9}	3.68×10^{-7}	8.12×10^{-5}	1.07×10^{-5}
f_9	1.12×10^{-8}	8.80×10^{-9}	4.02×10^{-8}	9.21×10^{-5}
f_{10}	4.72×10^{-9}	3.92×10^{-5}	2.20×10^{-4}	7.41×10^{-5}
f_{11}	4.72×10^{-9}	3.92×10^{-5}	2.20×10^{-4}	4.05×10^{-5}

Multimodal test functions with fixed dimensions

In the following experiments, the SMS algorithm is compared to GSA, PSO, DE and TPSO over a set of multidimensional functions with fixed dimensions which are widely used in the meta-heuristic literature. The functions used for the experiments are f_{12}, f_{13} and f_{14} which are presented in Table 6.10. The results in

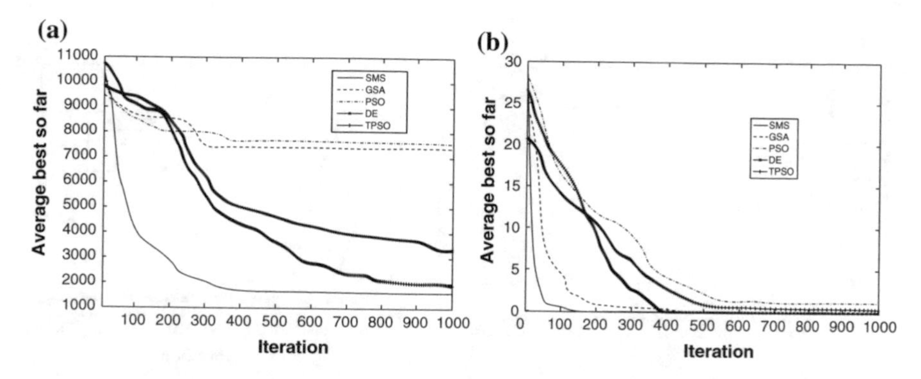

Fig. 6.5 Convergence rate comparison of PSO, GSA, DE, SMS and TPSO for minimization of **a** f_5 and **b** f_{11} considering $n = 30$

Table 6.5 Minimization results of benchmark functions in Table 6.10 with $n = 30$

		SMS	GSA	PSO	DE	TPSO
f_{12}	AB	**0.004361206**	0.051274735	0.020521847	0.006247895	0.008147895
	MB	0.004419241	0.051059414	0.020803912	0.004361206	0.003454528
	SD	0.004078875	0.016617355	0.021677285	8.7338E−15	6.37516E−15
f_{13}	AB	**−3.862782148**	−3.207627571	−3.122812884	−3.200286885	−3.311538343
	MB	−3.862782148	−3.222983851	−3.198877457	−3.200286885	−3.615938695
	SD	2.40793E−15	0.032397257	0.357126056	2.22045E−15	0.128463953
f_{14}	AB	**0**	0.00060678	1.07786E−11	4.45378E−31	0.036347329
	MB	3.82624E−12	0.000606077	0	4.93038E−32	0.002324632
	SD	2.93547E−11	0.000179458	0	1.0696E−30	0.032374213

Maximum number of iterations = 500
Bold elements represents the best values

Table 6.5 show that SMS, GSA, PSO, DE and TPSO have similar values in their performance. The evidence shows how meta-heuristic algorithms maintain a similar average performance when they face low-dimensional functions [54]. Figure 6.6 presents the convergence rate for the GSA, PSO, DE, SMS and TPSO algorithms considering functions f_{12} to f_{13}.

Test functions from the GECCO contest

The experimental set in Table 6.11 includes several representative functions that are used in the GECCO contest. Using such functions, the SMS algorithm is compared to GSA, PSO, DE and TPSO. The results have been averaged over 30 runs, reporting the performance indexes for each algorithm in Table 6.6. Likewise,

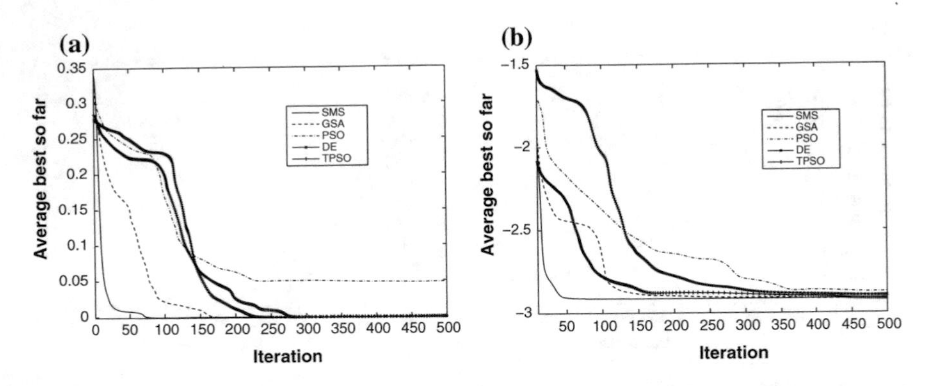

Fig. 6.6 Convergence rate comparison of PSO, GSA, DE, SMS and TPSO for minimization of **a** f_{12} and **b** f_{13}

Table 6.6 Minimization results of benchmark functions in Table 6.11 with $n = 30$

		SMS	GSA	PSO	DE	TPSO
f_{15}	AB	**−25.91760733**	57.15411412	134.3191481	183.6659439	−18.63859195
	MB	−29.92042882	57.38647154	133.1673936	186.723035	−21.73646913
	SD	23.85960437	14.20175945	68.4414947	38.0678428	12.54569285
f_{16}	AB	**−57.89720018**	−57.89605386	−40.5715691	−52.92227417	−50.437455071
	MB	−57.89733814	−57.89616319	−40.00561762	−53.25902658	−52.564574931
	SD	0.00077726	0.000841082	4.812411459	1.769678878	1.3446395342
f_{17}	AB	**184.7448285**	186.1082051	7540.2406	186.6192165	190.43463434
	MB	184.7424982	186.0937327	4831.581816	186.6285041	188.43649638
	SD	0.180957032	0.149285212	7101.466992	0.208918841	2.4340683134
f_{18}	AB	**−449.9936552**	2015.050538	18201.78495	−435.2972206	−410.37493561
	MB	−449.994798	1741.613119	18532.32174	−436.0279997	−429.46295713
	SD	0.005537064	1389.619208	6325.379751	2.880379023	1.4538493855
f_{19}	AB	**1213.421542**	22038.7467	30055.82961	43551.34835	1452.4364384
	MB	−181.0028277	21908.86945	26882.92621	42286.55626	1401.7493617
	SD	4050.267293	1770.050492	18048.55578	7505.414378	532.36343411
f_{20}	AB	**26975.80614**	66771.65533	44221.12187	58821.82993	29453.323822
	MB	24061.19301	65172.39992	44733.97226	60484.33588	28635.439023
	SD	10128.06919	12351.81976	16401.44428	9191.787618	4653.1269549
f_{21}	AB	**6526.690523**	23440.26883	23297.93668	26279.82607	7412.5361303
	MB	5716.886785	23427.99207	22854.63384	26645.28551	7012.4634613
	SD	2670.569217	2778.292017	5157.063617	2726.609286	745.37485621
f_{22}	AB	**965.8899213**	181742714.4	7385919478	284396.8728	1051.4348595
	MB	653.8161313	196616193.9	5789573763	287049.5324	1003.3448944
	SD	751.3821374	79542617.71	5799950322	66484.87261	894.43484589
f_{23}	AB	**18617.61336**	30808.74384	444370.5566	429178.9416	20654.323956
	MB	10932.4606	28009.57647	425696.8169	418480.2092	19434.343851
	SD	18224.4141	17834.72979	145508.9625	59342.54534	473.45938567
f_{24}	AB	**910.002925**	997.4123375	1026.555016	917.4176502	1017.3484548
	MB	910.0020976	999.1456735	1025.559417	917.3421337	993.34434754
	SD	0.004747964	19.08754967	57.01221298	0.456440816	45.343496836

Maximum number of iterations = 1000
Bold elements represents the best values

Table 6.7 p-values produced by Wilcoxon's test that compare SMS versus GSA, SMS versus PSO, SMS versus DE and SMS versus TPSO, for the "average best-so-far" (AB) values from Table 6.6

SMS versus	GSA	PSO	DE	TPSO
f_{15}	1.7344E−06	1.7344E−06	1.7344E−06	5.2334E−05
f_{16}	9.7110E−05	1.7344E−06	1.7344E−06	3.1181E−05
f_{17}	1.12654E−05	1.7344E−06	1.7344E−06	6.2292E−05
f_{18}	1.7344E−06	1.7344E−06	1.7344E−06	1.8938E−05
f_{19}	1.92092E−06	1.7344E−06	1.7344E−06	9.2757E−05
f_{20}	1.7344E−06	9.7110E−05	2.1264E−06	8.3559E−05
f_{21}	1.7344E−06	1.7344E−06	1.7344E−06	7.6302E−05
f_{22}	1.7344E−06	1.7344E−06	1.7344E−06	6.4821E−05
f_{23}	0.014795424	1.7344E−06	1.7344E−06	8.8351E−05
f_{24}	1.7344E−06	1.7344E−06	1.7344E−06	9.9453E−05

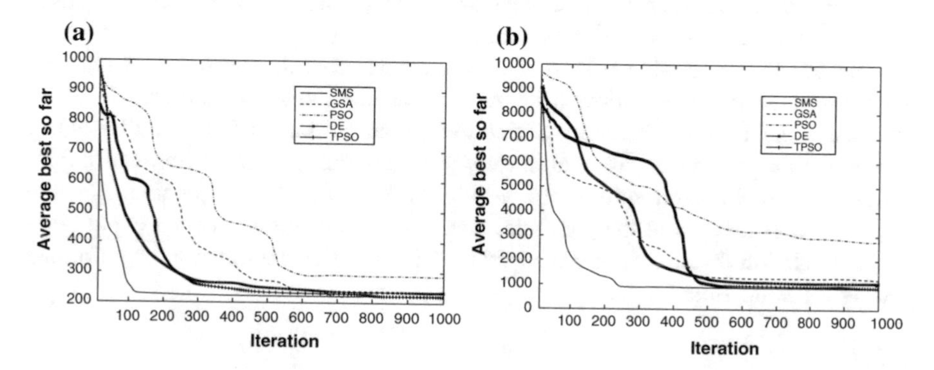

Fig. 6.7 Convergence rate comparison of PSO, GSA, DE, SMS and TPSO for minimization of **a** f_{17} and **b** f_{24}

p-values of the Wilcoxon signed-rank test of 30 independent executions are listed in Table 6.7. According to results of Table 6.6, it is evident that SMS yields much better solutions than other methods. The Wilcoxon test results in Table 6.7 provide information to statistically demonstrate that SMS has performed better than PSO, GSA, DE and TPSO. Figure 6.7 presents the convergence rate for the GSA, PSO, DE, SMS and TPSO algorithms, considering functions f_{17} to f_{24}.

6.5 Conclusions

In this chapter, a novel nature-inspired algorithm called as the States of Matter Search (SMS) has been introduced. The SMS algorithm is based on the simulation of the State of Matter phenomenon. In SMS, individuals emulate molecules which interact to each other by using evolutionary operations that are based on physical principles of the thermal-energy motion mechanism. The algorithm is devised by considering each state of matter at one different exploration–exploitation ratio. The evolutionary process is divided into three phases which emulate the three states of matter: gas, liquid and solid. At each state, molecules (individuals) exhibit different movement capacities. Beginning from the gas state (pure exploration), the algorithm modifies the intensities of exploration and exploitation until the solid state (pure exploitation) is reached. As a result, the approach can substantially improve the balance between exploration–exploitation, yet preserving the good search capabilities of an EA approach.

SMS has been experimentally tested considering a suite of 24 benchmark functions. The performance of SMS has been also compared to the following evolutionary algorithms: the Particle Swarm Optimization method (PSO) [17], the Gravitational Search Algorithm (GSA) [16], the Differential Evolution (DE) algorithm [13] and the PSO algorithm with a territorial diversity-preserving scheme (TPSO) [39]. Results have confirmed a high performance of the SMS method in terms of the solution quality for solving most of benchmark functions.

The SMS's remarkable performance is associated with two different reasons: (i) the defined operators allow a better particle distribution in the search space, increasing the algorithm's ability to find the global optima; and (ii) the division of the evolution process at different stages, provides different rates between exploration and exploitation during the evolution process. At the beginning, pure exploration is favored at the gas state, then a mild transition between exploration and exploitation features during liquid state. Finally, pure exploitation is performed during the solid state.

Appendix: List of Benchmark Functions

See Tables 6.8, 6.9, 6.10 and 6.11.

Table 6.8 Unimodal test functions

Test function	S	f_{opt}	n
$f_1(\mathbf{x}) = \sum_{i=1}^{n} x_i^2$	$[-100,100]^n$	0	30
$f_2(\mathbf{x}) = \max\{\lvert x_i \rvert, 1 \le i \le n\}$	$[-100,100]^n$	0	30
$f_3(\mathbf{x}) = \sum_{i=1}^{n-1} \left[100\left(x_{i+1} - x_i^2\right)^2 + (x_i - 1)^2\right]$	$[-30,30]^n$	0	30
$f_4(\mathbf{x}) = \sum_{i=1}^{n} i x_i^4 + rand(0,1)$	$[-1.28,1.28]^n$	0	30

Table 6.9 Multimodal test functions

Test function	S	f_{opt}	n
$f_5(\mathbf{x}) = 418.9829n + \sum_{i=1}^{n}\left(-x_i \sin\left(\sqrt{\lvert x_i\rvert}\right)\right)$	$[-500,500]^n$	0	30
$f_6(\mathbf{x}) = \sum_{i=1}^{50}\left(x_i^2 - 10\cos(2\pi x_i) + 10\right)$	$[-5.12,5.12]^n$	0	30
$f_7(\mathbf{x}) = \frac{1}{4000}\sum_{i=1}^{n} x_i^2 - \prod_{i=1}^{n}\cos\left(\frac{x_i}{\sqrt{i}}\right) + 1$	$[-600,600]^n$	0	30
$f_8(\mathbf{x}) = \frac{\pi}{n}\left\{10\sin(\pi y_1) + \sum_{i=1}^{n-1}(y_i - 1)^2\left[1 + 10\sin^2(\pi y_{i+1})\right] + (y_n - 1)^2\right\} + \sum_{i=1}^{n} u(x_i, 10, 100, 4)$ $y_i = 1 + \frac{(x_i+1)}{4}$ $u(x_i, a, k, m) = \begin{cases} k(x_i - a)^m & x_i > a \\ 0 & -a \le x_i \le a \\ k(-x_i - a)^m & x_i < a \end{cases}$	$[-50,50]^n$	0	30
$f_9(\mathbf{x}) = 0.1\left\{\sin^2(3\pi x_1) + \sum_{i=1}^{n}(x_i - 1)^2\left[1 + \sin^2(3\pi x_i + 1)\right] + (x_n - 1)^2\left[1 + \sin^2(2\pi x_n)\right]\right\} + \sum_{i=1}^{n} u(x_i, 5, 100, 4)$ where $u(x_i, a, k, m)$ is the same as f_8	$[-50,50]^n$	0	30
$f_{10}(\mathbf{x}) = \sum_{i=1}^{n} x_i^2 + \left(\sum_{i=1}^{n} 0.5ix_i\right)^2 + \left(\sum_{i=1}^{n} 0.5ix_i\right)^4$	$[-10,10]^n$	0	30
$f_{11}(\mathbf{x}) = 1 - \cos(2\pi\lVert x\rVert) + 0.1\lVert x\rVert$ where $\lVert x\rVert = \sqrt{\sum_{i=1}^{n} x_j^2}$	$[-100,100]^n$	0	30

Table 6.10 Multimodal test functions with fixed dimensions

Test function	S	f_{opt}	n
$f_{12}(\mathbf{x}) = \sum_{i=1}^{11}\left[a_i - \frac{x_i\left(b_i^2 + b_i x_2\right)}{b_i^2 + b_i x_3 + x_4}\right]^2$ $\mathbf{a} = [0.1957, 0.1947, 0.1735, 0.1600, 0.0844, 0.0627, 0.456, 0.0342, 0.0323, 0.0235, 0.0246]$ $\mathbf{b} = [0.25, 0.5, 1, 2, 4, 6, 8, 10, 12, 14, 16]$	$[-5,5]^n$	0.00030	4
$f_{13}(\mathbf{x}) = \sum_{i=1}^{4} c_i \exp\left(-\sum_{j=1}^{3} A_{ij}\left(x_j - P_{ij}\right)^2\right)$ $\mathbf{A} = \begin{bmatrix} 10 & 3 & 17 & 3.5 & 1.7 & 8 \\ 0.05 & 10 & 17 & 0.1 & 8 & 14 \\ 3 & 3.5 & 17 & 10 & 17 & 8 \\ 17 & 8 & 0.05 & 10 & 0.1 & 14 \end{bmatrix}$ $\mathbf{c} = [1, 1.2, 3, 3.2]$ $\mathbf{P} = \begin{bmatrix} 0.131 & 0.169 & 0.556 & 0.012 & 0.828 & 0.588 \\ 0.232 & 0.413 & 0.830 & 0.373 & 0.100 & 0.999 \\ 0.234 & 0.141 & 0.352 & 0.288 & 0.304 & 0.665 \\ 0.404 & 0.882 & 0.873 & 0.574 & 0.109 & 0.038 \end{bmatrix}$	$[0,1]^n$	-3.32	6
$f_{14}(\mathbf{x}) = (1.5 - x_1(1 - x_2))^2 + (2.25 - x_1(1 - x_2))^2 + (2.625 - x_1(1 - x_2))^2$	$[-4.5,4.5]^n$	0	2

Table 6.11 Set of representative GECCO functions

Test function	S	n	GECCO classification
$f_{15}(\mathbf{x}) = 10^6 \cdot z_1^2 + \sum_{i=2}^{n} z_i + f_{opt}$ $\mathbf{z} = T_{osz}(\mathbf{x} - \mathbf{x}^{opt})$ $T_{osz}: \mathbb{R}^n \to \mathbb{R}^n$, for any positive integer n, it maps element-wise $\mathbf{a} = T_{osz}(\mathbf{h})$, $\mathbf{a} = \{a_1, a_2, \ldots, a_n\}$, $\mathbf{h} = \{h_1, h_2, \ldots, h_n\}$ $a_i = \text{sign}(h_i)\exp(K + 0.049(\sin(c_1 K) + \sin(c_2 K)))$, where $K = \begin{cases} \log(h_i) & \text{if } h_i \neq 0 \\ 0 & \text{otherwise} \end{cases}, \text{sign}(h_i) = \begin{cases} -1 & \text{if } h_i < 0 \\ 0 & \text{if } h_i = 0, \\ 1 & \text{if } h_i > 0 \end{cases}$ $c_1 = \begin{cases} 10 & \text{if } h_i > 0 \\ 5.5 & \text{otherwise} \end{cases} \text{ and } c_2 = \begin{cases} 7.9 & \text{if } h_i > 0 \\ 3.1 & \text{otherwise} \end{cases}$	$[-5,5]^n$	30	GECCO2010 Discus function $f_{11}(\mathbf{x})$
$f_{16}(\mathbf{x}) = \sqrt{\sum_{i=1}^{n} \lvert z_i \rvert^{2+4\frac{i-1}{n-1}}} + f_{opt}$ $\mathbf{z} = \mathbf{x} - \mathbf{x}^{opt}$	$[-5,5]^n$	30	GECCO2010 Different powers function $f_{14}(\mathbf{x})$
$f_{17}(\mathbf{x}) = -\frac{1}{n}\sum_{i=1}^{n} z_i \sin\left(\sqrt{\lvert z_i \rvert}\right) + 4.189828872724339 + 100 f_{pen}\left(\frac{\mathbf{z}}{100}\right) + f_{opt}$ $\hat{\mathbf{x}} = 2 \times \mathbf{1}_{-}^{+} \otimes \mathbf{x}$ $\hat{z}_1 = \hat{x}_1, \hat{z}_{i+1} = \hat{x}_{i+1} + 0.25\left(\hat{x}_i - x_i^{opt}\right)$ for $i = 1, \ldots, n-1$ $\mathbf{z} = 100\left(\Lambda^{10}(\hat{\mathbf{z}} - \mathbf{x}^{opt}) + \mathbf{x}^{opt}\right)$ $f_{pen}: \mathbb{R}^n \to \mathbb{R}$, $a = f_{pen}(\mathbf{h})$, $\mathbf{h} = \{h_1, h_2, \ldots, h_n\}$ $a = 100\sum_{i=1}^{n} \max(0, \lvert h_i \rvert - 5)^2$ $\mathbf{1}_{-}^{+}$ is a n-dimensional vector with elements of -1 or 1 computed considering equal probability	$[-5,5]^n$	30	GECCO2010 Schwefel function $f_{20}(\mathbf{x})$
$f_{18}(\mathbf{x}) = \sum_{i=1}^{n} z_i^2 - 450$ $\mathbf{z} = \mathbf{x} - \mathbf{x}^{opt}$	$[-100,100]^n$	30	GECCO2005 Shifted sphere function $f_1(\mathbf{x})$
$f_{19}(\mathbf{x}) = \sum_{i=1}^{n}\left(\sum_{j=1}^{i} z_j\right)^2 - 450$ $\mathbf{z} = \mathbf{x} - \mathbf{x}^{opt}$	$[-100,100]^n$	30	GECCO2005 Shifted Schwefel's problem $f_2(\mathbf{x})$
$f_{20}(\mathbf{x}) = \left(\sum_{i=1}^{n}\left(\sum_{j=1}^{i} z_j\right)^2\right) \cdot (1 + 0.4\lvert N(0,1)\rvert) - 450$ $\mathbf{z} = \mathbf{x} - \mathbf{x}^{opt}$	$[-100,100]^n$	30	GECCO2005 Shifted Schwefel's problem 1.2 with noise in fitness $f_4(\mathbf{x})$
$f_{21}(\mathbf{x}) = \max\{\lvert \mathbf{A}\mathbf{x} - \mathbf{b}_i \rvert\} - 310$ $\mathbf{A}$ is a $n \times n$ matrix, $a_{i,j}$ are integer random numbers in the range $[-500,500]$, $\det(\mathbf{A}) \neq 0$ $\mathbf{b}_i = \mathbf{A}_i \cdot \mathbf{o}$ $\mathbf{A}_i$ is the ith row of $\mathbf{A}$ whereas $\mathbf{o}$ is a $n \times 1$ vector whose elements are random numbers in the range $[-100,100]$	$[-100,100]^n$	30	GECCO2005 Schwefel's problem 2.6 with global optimum on bounds $f_5(\mathbf{x})$
$f_{22}(\mathbf{x}) = \sum_{i=1}^{n}\left(100\left(z_i^2 - z_{i+1}\right)^2 + (z_i - 1)^2\right) + 390$ $\mathbf{z} = \mathbf{x} - \mathbf{x}^{opt}$	$[-100,100]^n$	30	GECCO2005 Shifted Rosenbrock's function $f_6(\mathbf{x})$
$f_{23}(\mathbf{x}) = \sum_{i=1}^{n}(A_i - B_i(\mathbf{x}))^2 - 460$ $A_i = \sum_{j=1}^{n}\left(a_{i,j}\sin\alpha_j + b_{i,j}\cos\alpha_j\right)$ $B_i(\mathbf{x}) = \sum_{i=1}^{n}\left(a_{i,j}\sin x_j + b_{i,j}\cos x_j\right)$ For $i = 1, \ldots, n$	$[-\pi,\pi]^n$	30	GECCO2005 Schwefel's problem 2.13 $f_{12}(\mathbf{x})$

(continued)

Table 6.11 (continued)

Test function	S	n	GECCO classification
$a_{i,j}$ and $b_{i,j}$ are integer random numbers in the range [−100,100], $\boldsymbol{\alpha} = [\alpha_1, \alpha_2, \ldots, \alpha_n]$, α_j are random numbers in the range $[-\pi,\pi]$			
$f_{24}(\mathbf{x}) = \sum_{i=1}^{10} \hat{F}_i(\mathbf{x} - \mathbf{x}_i^{opt})/\lambda_i$ $F_{1-2}(\mathbf{x})$ = Ackley's function $F_i(\mathbf{x}) = -20\exp\left(-0.2\sqrt{\frac{1}{D}\sum_{i=1}^{n} x_i^2}\right) - \exp\left(\frac{1}{D}\sum_{i=1}^{n}\cos(2\pi x_i)\right) + 20$ $F_{3-4}(\mathbf{x})$ = Rastringin's function $F_i(\mathbf{x}) = \sum_{i=1}^{n}\left(x_i^2 - 10\cos(2\pi x_i) + 10\right)$ $F_{5-6}(\mathbf{x})$ = Sphere function $F_i(\mathbf{x}) = \sum_{i=1}^{n} x_i^2$ $F_{7-8}(\mathbf{x})$ = Weierstrass function $F_i(x) = \sum_{i=1}^{n}\left(\sum_{k=0}^{kmax}\left[a^k\cos\left(2\pi b^k(x_i+0.5)\right)\right]\right) - n\sum_{k=0}^{kmax}\left[a^k\cos\left(2\pi b^k(x_i\cdot 0.5)\right)\right]$ $F_{9-10}(\mathbf{x})$ = Griewank's function $F_i(\mathbf{x}) = \sum_{i=1}^{n}\frac{x_i^2}{4000} - \prod_{i=1}^{n}\cos\left(\frac{x_i}{\sqrt{i}}\right) + 1$ $\hat{F}_i(\mathbf{z}) = F_i(\mathbf{z})/F_i^{\max}$. $F_i^{\max}$ is the maximum value of the particular function i $\boldsymbol{\lambda} = \left[\frac{10}{32}, \frac{5}{32}, 2, 1, \frac{10}{100}, \frac{5}{100}, 20, 10, \frac{10}{60}, \frac{5}{60}\right]$	$[-5,5]^n$	30	GECCO2005 Rotated version of hybrid composition function $f_{16}(\mathbf{x})$

The $\mathbf{x}^{opt}$ and f_{opt} values have been set to default values which have been obtained from the Matlab© implementation for GECCO competitions, as it is provided in [51]

References

1. Han, M.-F., Liao, S.-H., Chang, J.-Y., Lin, C.T.: Dynamic group-based differential evolution using a self-adaptive strategy for global optimization problems. Appl. Intell. https://doi.org/10.1007/s10489-012-0393-5
2. Pardalos Panos, M., Romeijn Edwin H., Tuy, H.: Recent developments and trends in global optimization. J. Comput. Appl. Math. **124**, 209–228 (2000)
3. Floudas, C., Akrotirianakis, I., Caratzoulas, S., Meyer, C., Kallrath, J.: Global optimization in the 21st century: advances and challenges. Comput. Chem. Eng. **29**(6), 1185–1202 (2005)
4. Ying, J., Ke-Cun, Z., Shao-Jian, Q.: A deterministic global optimization algorithm. Appl. Math. Comput. **185**(1), 382–387 (2007)
5. Georgieva, A., Jordanov, I.: Global optimization based on novel heuristics, low-discrepancy sequences and genetic algorithms. Eur. J. Oper. Res. **196**, 413–422 (2009)
6. Lera, D., Sergeyev, Y.: Lipschitz and Hölder global optimization using space-filling curves. Appl. Numer. Math. **60**(1–2), 115–129 (2010)
7. Fogel, L.J., Owens, A.J., Walsh, M.J.: Artificial Intelligence through Simulated Evolution. Wiley, Chichester, UK (1966)
8. De Jong, K.: Analysis of the behavior of a class of genetic adaptive systems. Ph.D. thesis, University of Michigan, Ann Arbor, MI (1975)
9. Koza, J.R.: Genetic programming: a paradigm for genetically breeding populations of computer programs to solve problems. Rep. No. STAN-CS-90-1314, Stanford University, CA (1990)
10. Holland, J.H.: Adaptation in Natural and Artificial Systems. University of Michigan Press, Ann Arbor, MI (1975)
11. Goldberg, D.E.: Genetic Algorithms in Search, Optimization and Machine Learning. Addison Wesley, Boston, MA (1989)
12. de Castro, L.N., Von Zuben, F.J.: Artificial immune systems: part I—basic theory and applications. Technical report, TR-DCA 01/99, December 1999

13. Storn, R., Price, K.: Differential evolution—a simple and efficient adaptive scheme for global optimisation over continuous spaces. Technical Report TR-95–012, ICSI, Berkeley, Calif (1995)
14. Kirkpatrick, S., Gelatt, C., Vecchi, M.: Optimization by simulated annealing. Science **220** (4598), 671–680 (1983)
15. İlker, B., Birbil, S., Shu-Cherng, F.: An electromagnetism-like mechanism for global optimization. J. Global Optim. **25**, 263–282 (2003)
16. Rashedia, E., Nezamabadi-pour, H., Saryazdi, S.: Filter modeling using gravitational search algorithm. Eng. Appl. Artif. Intell. **24**(1), 117–122 (2011)
17. Kennedy, J., Eberhart, R.: Particle swarm optimization. In: Proceedings of the 1995 IEEE International Conference on Neural Networks, vol. 4, pp. 1942–1948, December 1995
18. Dorigo, M., Maniezzo, V., Colorni, A.: Positive feedback as a search strategy. Technical Report No. 91-016, Politecnico di Milano (1991)
19. Tan, K.C., Chiam, S.C., Mamun, A.A., Goh, C.K.: Balancing exploration and exploitation with adaptive variation for evolutionary multi-objective optimization. Eur. J. Oper. Res. **197**, 701–713 (2009)
20. Chen, G., Low, C.P., Yang, Z.: Preserving and exploiting genetic diversity in evolutionary programming algorithms. IEEE Trans. Evol. Comput. **13**(3), 661–673 (2009)
21. Liu, S.-H., Mernik, M., Bryant, B.: To explore or to exploit: an entropy-driven approach for evolutionary algorithms. Int. J. Knowl. Based Intell. Eng. Syst. **13**(3), 185–206 (2009)
22. Alba, E., Dorronsoro, B.: The exploration/exploitation tradeoff in dynamic cellular genetic algorithms. IEEE Trans. Evol. Comput. **9**(3), 126–142 (2005)
23. Fister, I., Mernik, M., Filipič, B.: A hybrid self-adaptive evolutionary algorithm for marker optimization in the clothing industry. Appl. Soft Comput. **10**(2), 409–422 (2010)
24. Gong, W., Cai, Z., Jiang, L.: Enhancing the performance of differential evolution using orthogonal design method. Appl. Math. Comput. **206**(1), 56–69 (2008)
25. Joan-Arinyo, R., Luzon, M.V., Yeguas, E.: Parameter tuning of PBIL and CHC evolutionary algorithms applied to solve the root identification problem. Appl. Soft Comput. **11**(1), 754–767 (2011)
26. Mallipeddi, R., Suganthan, P.N., Pan, Q.K., Tasgetiren, M.F.: Differential evolution algorithm with ensemble of parameters and mutation strategies. Appl. Soft Comput. **11**(2), 1679–1696 (2011)
27. Sadegh, M., Reza, M., Palhang, M.: LADPSO: using fuzzy logic to conduct PSO algorithm. Appl. Intell. **37**(2), 290–304 (1012)
28. Yadav, P., Kumar, R., Panda, S.K., Chang, C.S.: An intelligent tuned harmony search algorithm for optimization. Inf. Sci. **196**(1), 47–72 (2012)
29. Khajehzadeh, M., Taha, M.R., El-Shafie, A., Eslami, M.: A modified gravitational search algorithm for slope stability analysis. Eng. Appl. Artif. Intell. **25**(8), 1589–1597 (2012)
30. Koumousis, V., Katsaras, C.P.: A saw-tooth genetic algorithm combining the effects of variable population size and reinitialization to enhance performance. IEEE Trans. Evol. Comput. **10**(1), 19–28 (2006)
31. Han, M.-F., Liao, S.-H., Chang, J.-Y., Lin, C.-T.: Dynamic group-based differential evolution using a self-adaptive strategy for global optimization problems. Appl. Intell. (2012). https://doi.org/10.1007/s10489-012-0393-5
32. Brest, J., Maučec, M.S.: Population size reduction for the differential evolution algorithm. Appl. Intell. **29**(3), 228–247 (2008)
33. Li, Y., Zeng, X.: Multi-population co-genetic algorithm with double chain-like agents structure for parallel global numerical optimization. Appl. Intell. **32**(3), 292–310 (2010)
34. Paenke, I., Jin, Y., Branke, J.: Balancing population- and individual-level adaptation in changing environments. Adapt. Behav. **17**(2), 153–174 (2009)
35. Araujo, L., Merelo, J.J.: Diversity through multiculturality: assessing migrant choice policies in an island model. IEEE Trans. Evol. Comput. **15**(4), 456–468 (2011)
36. Gao, H., Xu, W.: Particle swarm algorithm with hybrid mutation strategy. Appl. Soft Comput. **11**(8), 5129–5142 (2011)

37. Jia, D., Zheng, G., Khan, M.K. (2011). An effective memetic differential evolution algorithm based on chaotic local search. Inf. Sci. **181**(15), 3175–3187
38. Lozano, M., Herrera, F., Cano, J.R.: Replacement strategies to preserve useful diversity in steady-state genetic algorithms. Inf. Sci. **178**(23), 4421–4433 (2008)
39. Ostadmohammadi, B., Mirzabeygi, P., Panahi, M.: An improved PSO algorithm with a territorial diversity-preserving scheme and enhanced exploration–exploitation balance. Swarm Evol. Comput. (In Press)
40. Yang, G.-P., Liu, S.-Y., Zhang, J.-K., Feng, Q.-X.: Control and synchronization of chaotic systems by an improved biogeography-based optimization algorithm. Appl. Intell. https://doi.org/10.1007/s10489-012-0398-0
41. Hasanzadeh, M., Meybodi, M.R., Ebadzadeh, M.M.: Adaptive cooperative particle swarm optimizer. Appl. Intell. https://doi.org/10.1007/s10489-012-0420-6
42. Aribarg, T., Supratid, S., Lursinsap, C.: Optimizing the modified fuzzy ant-miner for efficient medical diagnosis. Appl. Intell. **37**(3), 357–376 (2012)
43. Fernandes, C.M., Laredo, J.L.J., Rosa, A.C., Merelo, J.J.: The sandpile mutation genetic algorithm: an investigation on the working mechanisms of a diversity-oriented and self-organized mutation operator for non-stationary functions. Appl. Intell. https://doi.org/10.1007/s10489-012-0413-5
44. Gwak, J., Sim, K.M.: A novel method for coevolving PS-optimizing negotiation strategies using improved diversity controlling EDAs. Appl. Intell. **38**(3), 384–417 (2013)
45. Cheshmehgaz, H.R., Desa, M.I., Wibowo, A.: Effective local evolutionary searches distributed on an island model solving bi-objective optimization problems. Appl. Intell. **38**(3), 331–356 (2013)
46. Cuevas, E., González, M.: Multi-circle detection on images inspired by collective animal behavior. Appl. Intell. https://doi.org/10.1007/s10489-012-0396-2
47. Adra, S.F., Fleming, P.J.: Diversity management in evolutionary many-objective optimization. IEEE Trans. Evol. Comput. **15**(2), 183–195 (2011)
48. Črepineš, M., Liu, S.H., Mernik, M.: Exploration and exploitation in evolutionary algorithms: a survey. ACM Comput. Surv. **1**(1), 1–33 (2011)
49. Ceruti, G., Rubin, H.: Infodynamics: analogical analysis of states of matter and information. Inf. Sci. **177**, 969–987 (2007)
50. Chowdhury, D., Stauffer, D.: Principles of Equilibrium Statistical Mechanics, 1st edn. Wiley-VCH, Germany (2000)
51. Betts, D.S., Turner, R.E.: Introductory Statistical Mechanics, 1st edn. Addison Wesley, Boston (1992)
52. Cengel, Y.A., Boles, M.A.: Thermodynamics: An Engineering Approach, 5th edn. McGraw-Hill, USA (2005)
53. Bueche, F., Hecht, E.: Schaum's Outline of College Physics, 11th edn. McGraw-Hill, USA (2012)
54. Piotrowski, A.P., Napiorkowski, J.J., Kiczko, A.: Differential evolution algorithm with separated groups for multi-dimensional optimization problems. Eur. J. Oper. Res. **216**(1), 33–46 (2012)
55. Mariani, V.C., Luvizotto, L.G.J., Guerra, F.A., dos Santos Coelho, L.: A hybrid shuffled complex evolution approach based on differential evolution for unconstrained optimization. Appl. Math. Comput. **217**(12), 5822–5829 (2011)
56. Yao, X., Liu, Y., Lin, G.: Evolutionary programming made faster. IEEE Trans. Evol. Comput. **3**(2), 82–102 (1999)
57. Moré, J.J., Garbow, B.S., Hillstrom, K.E.: Testing unconstrained optimization software. ACM Trans. Math. Softw. **7**(1), 17–41 (1981)
58. Tsoulos, I.G.: Modifications of real code genetic algorithm for global optimization. Appl. Math. Comput. **203**(2), 598–607 (2008)
59. Black-Box Optimization Benchmarking (BBOB) 2010. In: 2nd GECCO Workshop for Real-Parameter Optimization. http://coco.gforge.inria.fr/doku.php?id=bbob-2010

60. Hedar, A.-R., Ali, A.F.: Tabu search with multi-level neighborhood structures for high dimensional problems. Appl. Intell. **37**(2), 189–206 (2012)
61. Vafashoar, R., Meybodi, M.R., Momeni Azandaryani, A.H.: CLA-DE: a hybrid model based on cellular learning automata for numerical optimization. Appl. Intell. **36**(3), 735–748 (2012)
62. Garcia, S., Molina, D., Lozano, M., Herrera, F.: A study on the use of non-parametric tests for analyzing the evolutionary algorithms' behaviour: a case study on the CEC '2005, Special session on real parameter optimization. J. Heurist (2008). https://doi.org/10.1007/s10732-008-9080-4
63. Shilane, D., Martikainen, J., Dudoit, S., Ovaska, S.: A general framework for statistical performance comparison of evolutionary computation algorithms. Inf. Sci. **178**, 2870–2879 (2008)

Chapter 7
Multimodal States of Matter Search

The idea in multi-modal optimization is to detect multiple global and local optima as possible in only one run. Identifying several solutions is particularly important for some problems because the best solution could not be applicable due to different practical limitations. The States of Matter Search (SMS) is a metaheuristic technique. Even though SMS is efficient in finding the global optimum, it misses in providing various solutions by using an only single run. Under this condition, a new version called the Multi-modal States of Matter Search (MSMS) has been proposed. In this chapter, the performance of MSMS to optimize multi-modal problems is analyzed. In MSMS, the original SMS is improved with new multimodal features such as: (I) a memory model to register high-quality local optima and their distance to other presumably promising solutions; (II) the alteration of the original SMS search strategy to quicken the identification of new local minima; and (III) the addition of a eliminating process at the end of each phase to exclude duplicated memory individuals.

7.1 Introduction

Optimization is a field with applications in many areas of science, engineering, economics and others, where mathematical modeling is used [1]. In optimization, the objective is to identify an acceptable solution to an objective function defined over a certain domain. Optimization methods are usually broadly divided into classic and stochastic [2]. Classic techniques present several difficulties in solving optimization problems [3], since they impose different restrictions to the optimization formulations to be solved. On the other hand, stochastic methods are usually faster in locating a global optimum [4]. Furthermore, stochastic algorithms adapt easily to black-box problem definitions and ill-behaved functions, whereas classical methods require the existence of some technical constraints about the problem and its analytical properties (such as Lipschitz continuity) [5].

E. Cuevas et al., *Advances in Metaheuristics Algorithms: Methods and Applications*, Studies in Computational Intelligence 775,
https://doi.org/10.1007/978-3-319-89309-9_7

Evolutionary algorithms (EA) symbolize the most prominent members of the stochastic methods. They are designed considering the combination of deterministic rules and random process simulating several natural systems. Such systems comprise evolutionary phenomena such as the Evolutionary Strategies (ES) proposed by Fogel et al. [6], Schwefel [7], and Koza [8], the Genetic Algorithm (GA) introduced by Holland [9] and Goldberg [10], the Artificial Immune System considered by De Castro and Von Zuben [11] and the Differential Evolution Algorithm (DE) introduced by Price and Storn [12]. Alternatively to these methods, other algorithms based on physical phenomena have appeared in the literature. They consider the Simulated Annealing (SA) introduced by Kirkpatrick et al. [13], the Electromagnetism-like Algorithm proposed by İlker et al. [14] and the Gravitational Search Algorithm proposed by Rashedia et al. [15]. Additionally, there exist other techniques based on the emulation of animal-behavior systems such as the Particle Swarm Optimization (PSO) algorithm introduced by Kennedy and Eberhart [16] and the Ant Colony Optimization (ACO) algorithm proposed by Dorigo et al. [17].

Most of the research on EA is concentrated in the localization of only one global optimum [18]. Despite its best properties, the use of the global optimum as the solution to an optimization problem can be considered unfeasible or excessively expensive, restricting their selection in several real-world applications. Hence, in practice, it is better to detect not only the best solution but also as many local solutions as possible. In these conditions, one local optimum with a satisfactory quality and a reasonable cost can be preferred instead of a costly global optimum that offers only a marginal superior performance [19]. Multimodal optimization is defined as the process of detecting all the global optima and multiple local optima for a particular optimization formulation.

Each EA method requires the balance between the exploration and exploitation of the search space [20]. The process of examining new possible solutions in the entire search space is known an exploration. In contrast, exploitation is the action of locally improving already visited solutions within a small region around them. The use of only exploration deteriorates the accuracy of the optimization process but enhances its potential to find new promising individuals. On the contrary, the fact of exclusively using exploitation permits to improve significantly existent solutions. However, it unfavorably conducts to local optimal solutions.

EA perform well for locating a single global optimum but fail to provide multiple solutions [18]. The process of detecting all the optima in only one execution is harder than global optimization. Detection and maintenance of multiple solutions are two fundamental processes in EAs for facing multi-modal optimization problems, since EAs have been originally conceived to detect only a global solution. Several techniques that commonly referred to as niching approaches have been incorporated into the original EAs to make them suitable for solving multimodal optimization problems [21]. Niching is a method for finding and preserving multiple stable subpopulations to avoid the convergence in a single point. The main objective of a niching technique is to conserve the individual diversity. Therefore, most of the techniques for multimodal optimization are typically based on diversity maintenance methods borrowed from other computational or biological domains [22].

Several niching methods have been proposed in the literature including crowding [23], fitness sharing [24], clearing [25] and speciation [26].

The concept of crowding was firstly introduced by De Jong in 1975 [24, 27] to preserve the diversity of the population. In crowding, each offspring is compared to other solutions which are randomly extracted from the current population. Then, the most similar individual is replaced by the offspring that maintains a superior solution. Several multimodal algorithms have been conceived in the literature considering the crowding principles. Some examples include different evolutionary methods such as Differential Evolution (DE) [28, 29], Genetic Algorithms (GA) [30], Gravitational Search Algorithm (GSA) [31] and Particle Swarm Optimization (PSO) [32].

Fitness sharing [24] is one of the most well-known methods for producing subpopulations of similar individuals. Fitness sharing considers that a fitness value of a determined location needs to be distributed by individuals which occupy similar positions. Fitness sharing in EAs is implemented by degrading the fitness value of an individual according to the number of similar individuals present in the population. The degradation of a fitness value is controlled by two operations: a similarity function and a sharing model. The similarity function measures the distance between two individuals in order to evaluate their affinity. The purpose of the sharing model is to take the similarity between two individuals and return the degree to which they must degrade their fitness values since they are considered as the same species. Considering different variants of similarity function and a sharing model, several multimodal approaches have been suggested including the examples of the Fitness Sharing Differential Evolution (SDE) [28, 29], the Fitness Euclidean-distance ratio method [33] and the Information Sharing Artificial Bee Colony algorithm [34].

Clearing [25], different to fitness sharing, considers the best individuals of the sub-populations and removes the remaining population members. Under this procedure, the algorithm first sorts the individuals in descending order on their fitness values. Then, at each iteration, it picks one individual from the top and removes all other individuals that hold a worse fitness within a specified clearing radius. This process will be repeated until all the individuals in the population are either selected or removed. As a result, clearing eliminates similar individuals and maintains the diversity among the selected individuals. Depending on the definition of the clearing radius, new multimodal methods have been introduced; some examples include the Clearing Procedure (CP) [25] and the Topographical Clearing Differential Evolution [35].

Speciation-based niching techniques require the partition of the complete population into different sub-populations considered as species. Existing speciation methods mainly differ in the way they determine whether two individuals are of the same species. Therefore, they can be divided into distance-based and topology-based. Distance-based methods rely on the idea that closer individuals are more likely to be of the same species. Typically, a fixed radius is defined. Under such conditions, two individuals are identified of the same species if their distance is below such radius. Some examples of distance-based algorithms consider the

Elitist-population strategy (AEGA) [36], the Differential Evolution with Self-adaptive strategy [37] and the Distance-based PSO [38]. On the other hand, topology-based methods rely on the intuition that two individuals should be of different species if there exists a determined location configuration between them on the fitness landscape. This methodology makes weaker assumptions about the fitness landscape than the distance-based one, and it is able to form species of different sizes and shapes. Some multimodal algorithms of this category include the Gaussian Classifier-based EA [39], the Ensemble Speciation DE [40] and the History-based Topological speciation [26].

However, most of the niching methods have difficulties that need to be overcome before they can be successfully employed for multimodal purposes. Some known problems involve the setting of some niching parameters, the storage of discovered solutions during the execution, the high computational cost and the low scalability when the number of dimensions is high. Another significant problem represents the fact that such methods are devised for extending the search capacities of popular EA such as GA, DE and PSO which fail in finding a balance between exploration and exploitation, mainly for multimodal functions [41]. Furthermore, they do not explore the whole region effectively and often suffer premature convergence or loss of diversity.

As alternative to niching approaches, other bio-inspired algorithms with multimodal capacities have been devised. These methods consider as fundament our scientific perception of biological phenomena, which can be abstracted as multimodal optimization processes. Some examples of these methods are the Clonal Selection Algorithm [42] and the Artificial Immune Network (AiNet) [43, 44], the Collective Animal Behavior algorithm (CAB) [45], the Runner-root method [46], the Multimodal Gravitational Search algorithm (MGSA) [31] and the Region-Based Memetic method (RM) [47]. Such approaches employ operators and structures which support the finding and maintaining of multiple-solutions.

To detect multiple-solutions, multimodal methods demands an adequate level of exploration of the search space. Under such conditions, most of the local and global locations can be successfully found [48, 49]. Furthermore, an effective multimodal optimization method should present not only a good exploration behavior but also a good exploitative performance. This fact is of particular importance at the end of the optimization process since it must guarantee the convergence in different optima of the search space. Therefore, the ability of an EA to find multiple solutions depends on its capacity to reach an adequate balance between the exploitation of found-so-far elements and the exploration of the search space [31]. Up to date, the exploration–exploitation dilemma has been an unsolved issue within the framework of EA.

Recently, a novel nature-inspired algorithm called the States of Matter Search (SMS) [50] has been introduced in order to solve complex optimization formulations. The SMS algorithm is based on the natural phenomenon of the states of matter. In SMS, candidate solutions emulate molecules that interact with each other by using evolutionary operators extracted from the physical principles of the thermal-energy motion mechanism. Such operations allow the increase of the population diversity and avoid the concentration of particles (search points) in local minima. SMS considers the use of evolutionary elements with a control procedure that adjusts the configuration of each operation during the search strategy. In general, conventional EA algorithms improve its exploration–exploitation rate by the incorporation of additional mechanisms. Different to these methods, the SMS approach permits to adjust the exploration–exploitation compromise as a consequence of its search strategy. In SMS, each state of matter represents a particular exploration–exploitation degree. Therefore, the complete optimization process is separated into three periods which emulate the different states of matter: gas, liquid and solid. For each state transition, molecules (individuals) manifest distinct behaviors. Starting from the gas state (with only exploration), the method changes the levels of exploration and exploitation until the last period (solid state with only exploitation) is attained. As a result, SMS can substantially improve the balance between exploration–exploitation, increasing its multimodal search capabilities. These capacities make SMS a suitable approach for solving complex optimization problems such as template matching in image processing [51] and energy transmission in power systems [52].

This chapter introduces a new multimodal optimization algorithm called the Multi-modal States of Matter Search (MSMS). The method combines the SMS approach with a memory system that allows an effective storage of promising local optima according to their fitness quality and the distance to other potential solutions. The original SMS search strategy is mainly conducted by the best individual found so-far (global optimum). To increase the speed of local minima detection, the original evolution process is altered to be controlled by solutions that are included in the memory system. During each state, molecules (individuals) that exhibit a good fitness quality are incorporated within the memory. In the storage process, several individuals could refer to the same local optimum, because of that a refinement procedure is also included at the end of each state to exclude multiple memory members. To examine the performance of the MSMS multimodal approach, it is compared to different state-of-the-art multimodal optimization methods over a set of 21 multimodal problems. Simulation results confirm that our strategy is able to maintain a better and more regular performance than its counterparts for most of test problems considering a low computational cost.

The chapter is organized as follows: Sect. 7.2 explains the States of Matter Search (SMS) while Sect. 7.3 presents the MSMS method. Section 7.4 exposes the experimental results. Finally, Sect. 7.5 establishes some concluding remarks.

7.2 Original States of Matter Search (SMS)

In general, the matter maintains three phases known as gas, liquid and solid. In each state, the forces among the molecules control the permissible distance ρ that the particles can move with each other. In the gas state, particles can displace a large allowable distance. On the other hand, in the liquid state, this distance is partially reduced. Finally, in the solid state, only a small movement (i.e. vibration) is allowed.

In SMS, candidate solutions are considered as molecules whose positions on a multidimensional space are modified as the algorithm evolves. The movement of such candidate solutions emulates the physical state transition process experimented by molecules as a consequence of thermal-energy laws.

7.2.1 Definition of Operators

In the SMS operation, a population $\mathbf{P}^k$ $(\{\mathbf{p}_1^k, \mathbf{p}_2^k, \ldots, \mathbf{p}_N^k\})$ of N molecules (individuals) is evolved from the initial point (k = 1) to a total *gen* number iterations ($k = gen$). Each molecule $\mathbf{p}_i^k$ ($i \in [1, \ldots, N]$) represents an n-dimensional vector $\left\{p_{i,1}^k, p_{i,2}^k, \ldots, p_{i,n}^k\right\}$ where each dimension corresponds to a decision variable of the optimization problem to be solved. The quality of each molecule $\mathbf{p}_i^k$ (candidate solution) is evaluated by using an objective function $f\left(\mathbf{p}_i^k\right)$ whose final result represents the fitness value of $\mathbf{p}_i^k$. As the optimization process evolves, the best individual $\mathbf{p}^{best}$ seen so-far is conserved, since it represents the current best available solution. During the evolution process, SMS applies three operators over the population $\mathbf{P}^k$. In the next paragraphs, such operators are described.

7.2.1.1 Direction Vector

This operator mimics the way in which molecules change their positions according to the thermic-energy laws. Therefore, for each molecule $\mathbf{p}_i^k$, it is assigned an n-dimensional direction vector $\mathbf{d}_i^k$ ($\mathbf{d}_i^k = \left\{d_{i,1}^k, d_{i,2}^k, \ldots, d_{i,n}^k\right\}$). Initially, all the direction vectors are randomly chosen within the range of $[-1,1]$. As SMS evolves, each direction vector i is iteratively computed considering the following model:

$$\mathbf{d}_i^{k+1} = \mathbf{d}_i^k \cdot \left(1 - \frac{k}{gen}\right) \cdot 0.5 + \mathbf{a}_i, \tag{7.1}$$

where $\mathbf{a}_i$ represents the attraction unitary vector calculated as $\mathbf{a}_i = (\mathbf{p}^{best} - \mathbf{p}_i^k)/\left\|\mathbf{p}^{best} - \mathbf{p}_i^k\right\|$. Then, in all states $(0.0 \leq \rho \leq 1)$, the new position for each molecule i is updated as follows:

$$p_{i,j}^{k+1} = p_{i,j}^k + v_{i,j}^k \cdot \text{rand}(0,1) \cdot \rho \cdot (b_j^{high} - b_j^{low}) \tag{7.2}$$

where $v_{i,j}^{k}$ represents the velocity at time k for the particle i in its dimension j. It can be calculated by the following formulation:

$$v_{i,j}^{k} = d_{i,j}^{k} \cdot \frac{\sum_{j=1}^{n} (b_j^{high} - b_j^{low})}{n} \cdot \beta \tag{7.3}$$

where b_j^{low} and b_j^{high} are the low j parameter bound and the upper j parameter bound respectively, whereas $\beta \in [0,1]$ is a tuning parameter.

7.2.1.2 Collision Operation

This operation simulates the collisions verified by two molecules $\mathbf{p}_i$ and $\mathbf{p}_q$ when they interact to each other. This operator is considered if the condition $\|\mathbf{p}_i - \mathbf{p}_q\| < r$ is satisfied, where r represents collision radius. If a collision occurs, the direction vector for each particle is modified by interchanging their respective direction vectors as follows:

$$\mathbf{d}_i = \mathbf{d}_q \text{ and } \mathbf{d}_q = \mathbf{d}_i \tag{7.4}$$

The collision radius is calculated by:

$$r = \frac{\sum_{j=1}^{n} (b_j^{high} - b_j^{low})}{n} \cdot \alpha \tag{7.5}$$

where $\alpha \in [0,1]$.

7.2.1.3 Random Positions

In SMS, the random behavior of molecules is emulated as a reinitialized process. In the operation, particles are randomly modified considering a probabilistic criterion dependent on a threshold parameter $H \in [0,1]$. The operator can be formulated as follows:

$$p_{i,j}^{k+1} = \begin{cases} b_j^{low} + \text{rand}(0,1) \cdot (b_j^{high} - b_j^{low}) & \text{with probability } H \\ p_{i,j}^{k+1} & \text{with probability } (1 - H) \end{cases} \tag{7.6}$$

where $i \in \{1, \ldots, N_p\}$ and $j \in \{1, \ldots, n\}$.

7.2.2 General Procedure

The ability of an EA to find a global optimal solution depends on its capacity to find a good balance between the exploitation of found-so-far elements and the exploration of new solutions [31]. So far, the exploration–exploitation dilemma has not been satisfactorily solved within the framework of EA. Under such circumstances, in order to balance exploration–exploitation, SMS divides the evolution process in three stages: Gas, liquid and solid. Such states are sequentially applied according to the following rate 50:40:10, respectively. In each state, the algorithm presents a different exploration–exploitation ratio. It goes from pure exploration at gas state to pure exploitation at the solid state, considering an intermediate liquid state where a combination of both is achieved. The duration of each state proposed by SMS has been experimentally determined [50] so that its exploration–exploitation equilibrium permits to successfully solve most of the optimization problems known in the literature for testing the performance of an evolutionary computation technique. The duration of each state can be also modified to alter the exploration–exploitation rate when a particular optimization problem demands a different exploration–exploitation proportion [51, 52].

In SMS, the three operators are applied at each state. However, depending on which state is referred, they are employed considering a different parameter configuration. Table 7.1 presents the parameter setting of each state. More details of SMS can be seen in [51] and [50].

7.2.3 Parameter Analysis

Under SMS, the optimization process is divided into three stages which represent different exploration–exploitation behaviors. In order to define a determined exploration–exploitation behavior for each stage, a proper configuration of the parameter set is required. The SMS algorithm has been designed considering four parameters (ρ, β, α and H) and the duration for each stage. From all these specifications, only ρ can be manipulated by the user to obtain a particular performance. The remaining data β, α, H and the duration of each process are set to specific constants that have been already defined by the original algorithm. Such constants represent, according to [50], the optimal values which assure the appropriate

Table 7.1 SMS control parameter settings

State	Gas	Liquid	Solid
ρ	$\in [0.8,1]$	$\in [0.3,0.6]$	$\in [0,0.1]$
β	0.8	0.4	0.1
α	0.8	0.2	0
H	0.9	0.2	0
k	50% (gen)	40% (gen)	10% (gen)

performance of SMS at each stage. Therefore, in this chapter β, α and H and the duration of each stage are configured according to the specific values proposed by SMS.

In SMS, new individuals are generated through the modification of already existent individuals. Under such circumstances, the parameter ρ regulates the magnitude of this modification. Considering that ρ is the only configurable parameter, it must be adequately calculated in order to obtain the best possible performance of SMS for a specific optimization context. Since SMS divides the optimization process in three phases, the identification of ρ actually involves the definition of three different parameters: ρ_1 for the gas state, ρ_2 for the liquid state and ρ_3 for the solid state.

To study the effect of parameters ρ_1, ρ_2 and ρ_3 over the performance of SMS, an experiment based on Design of Experiments (DoE) [53] has been conducted. DoE is an efficient methodology for retrieving optimal parameter settings in a way that the obtained information can be examined to produce valid and objective deductions. This methodology has been widely adopted by the EA community [54–63] for parameter tuning of several evolutionary computation algorithms such as Evolutionary Strategies [64, 65], Genetic Programming [66], Genetic Algorithms [67, 68], PSO [55], DE [62], ACO [69] and SA [70].

Under the DoE methodology, the process variables or factors $(\theta_1, \theta_2, \ldots, \theta_P)$ is randomly modified in order to observe the effect over the response variable (Y). The main assumption to apply DoE to EA is the following: the factors of a DoE are the parameters for an EA whereas the response is defined as the quality of the results of the EA (e.g. the best fitness value, the average fitness value, or its convergence ratio).

One important technique in DoE is the Response Surface Modeling (RSM) [55, 67, 68, 71] which permits to find the optimal factors through the maximization of the response in the process. In RMS, it is common to begin with a model of the black box type, with multiple discrete or continuous input parameters that can be adjusted, as well as the measured output response. It is assumed that output responses are continuous values. Then, experimental data are used to derive an empirical (approximation) model linking the outputs and inputs. The most common empirical model considered by RMS is the linear form. This model, considering as example three factors $(\theta_1, \theta_2, \theta_3)$, is presented in the following equation:

$$Y = X_1\theta_1 + X_2\theta_2 + X_3\theta_3 + X_4\theta_1^2 + X_5\theta_2^2 + X_6\theta_3^2 + X_7\theta_1\theta_2 + X_8\theta_1\theta_3 + X_9\theta_2\theta_3 + K \tag{7.7}$$

where $X_1, \ldots, X_9$ represent the importance levels corresponding to each parameter $(\theta_1, \theta_2, \theta_3)$ or parameter combination $(\theta_1\theta_2, \theta_1\theta_3, \theta_2\theta_3)$. K symbolizes an importance level whose value does not have relation with any parameter. For each variable from $X_1, \ldots, X_9$, RSM calculates a statistics p-value that evaluates its respective significance in the process. If the p-value is greater than 0.05 (for a 95% of significance), the parameter or parameter combination influence is neglected.

Once obtained Eq. 7.9, it is analyzed, by using calculus, the importance levels ($X_1, \ldots, X_9$) which produce that the random variable Y reaches its minimum value. Therefore, the optimal values of θ_1, θ_2 and θ_3 are obtained by differentiating (X) with respect to each factor in turn, setting each derivate equal zero ($\partial Y/\partial\theta_1 = 0$, $\partial Y/\partial\theta_2 = 0$, $\partial Y/\partial\theta_3 = 0$) and solving the resulting system of equations. A detailed description of the RSM method and the DoE methodology is provided in Bartz-Beielstein [55], Petrovski et al. [25].

7.2.4 Parameter Tuning Results

In order to determine the optimal values $\hat{\rho}_1$, $\hat{\rho}_2$ and $\hat{\rho}_3$ of ρ_1, ρ_2 and ρ_3, an experiment based on DoE methodology has been conducted. The experiment considers the optimization of the five multimodal functions that are shown in Table 7.2. In such functions, n represents the dimension of function, f_{opt} is the minimum value of the function and S is a subset of R^n. The optimum location ($\mathbf{x}_{opt}$) for the functions fall into $[0]^n$, except for F_5 with $\mathbf{x}_{opt}$ falling into $[1]^n$.

In the experiment, an RSM analysis is independently conducted for each function. In each RSM application, 100 different executions of SMS over the same function are considered. At each execution, the parameters ρ_1, ρ_2 and ρ_3 within their valid intervals ($\rho_1 \in [0.8,1]$, $\rho_2 \in [0.3,0.6]$ and $\rho_3 \in [0.0,0.1]$) are randomly modified in order to observe the effect into the final best fitness value (Y). In the executions, the maximum number of iterations is set to 1000 whereas the number of individuals N has been configured to 50. For the sake of space, only the process of RSM for F_5 is presented. Table 7.3 present the RSM results after analyzing the data produced for the 100 different executions of SMS over function F_5. Using such data, the empirical linear model is built using the following coefficients:

$$\begin{aligned} Y_{f_5} = & -1.71\rho_1 - 2.08\rho_2 + -1.88\rho_3 + 0.65\rho_1^2 + 1.20\rho_2^2 + 1.35\rho_3^2 \\ & + 0.83\rho_1\rho_2 + 1.16\rho_1\rho_3 + 1.18\rho_2\rho_3 - 3.41 \end{aligned} \tag{7.8}$$

Since all p-values of the coefficients are less than 0.05, all coefficients must be considered as significant, appearing in the final expression of Eq. 7.8. By considering such data, the optimal parameters $\hat{\rho}_1^{F_5}$, $\hat{\rho}_2^{F_5}$ and $\hat{\rho}_3^{F_5}$ for function F_5 are calculated by differentiating (X2) with respect to each parameter in turn, setting each derivate equal zero ($\partial Y_{F_5}/\partial\rho_1 = 0, \partial Y_{F_5}/\partial\rho_2 = 0, \partial Y_{F_5}/\partial\rho_3 = 0$) and solving the resulting system of equations. After achieving the complete procedure, it has been found that $\hat{\rho}_1^{F_5} = 0.92$, $\hat{\rho}_2^{F_5} = 0.51$ and $\hat{\rho}_3^{F_5} = 0.078$.

The optimal parameters for functions F_1–F_4 are also computed following the same procedure for F_5. After the definition of the optimal parameters $\left(\hat{\rho}_z^{F_w}; z \in 1,2,3; w \in 1,\ldots,5\right)$ for all functions, the final optimal value of each parameter $\hat{\rho}_z$ is calculated as the averaged optimal values produced for the five

Table 7.2 Multimodal test functions

Test function	S	f_{opt}
$F_1(\mathbf{x}) = \sum_{i=1}^{n} -x_i \sin\left(\sqrt{\lvert x_i \rvert}\right)$	$[-500,500]^n$	$-418.98*n$
$F_2(\mathbf{x}) = \sum_{i=1}^{n} \left[x_i^2 - 10\cos(2\pi x_i) + 10\right]$	$[-5.12,5.12]^n$	0
$F_3(\mathbf{x}) = -20\exp\left(-0.2\sqrt{\frac{1}{n}\sum_{i=1}^{n} x_i^2}\right) - \exp\left(\frac{1}{n}\sum_{i=1}^{n}\cos(2\pi x_i)\right) + 20$	$[-32,32]^n$	0
$F_4(\mathbf{x}) = \frac{1}{4000}\sum_{i=1}^{n} x_i^2 - \prod_{i=1}^{n}\cos\left(\frac{x_i}{\sqrt{i}}\right) + 1$	$[-600,600]^n$	0
$F_5(\mathbf{x}) = 0.1\left\{\sin^2(3\pi x_1) + \sum_{i=1}^{n}(x_i - 1)^2\left[1 + \sin^2(3\pi x_i + 1)\right] + (x_n - 1)^2\left[1 + \sin^2(2\pi x_n)\right]\right\} + \sum_{i=1}^{n} u(x_i, 5, 100, 4)$	$[-50,50]^n$	0

Table 7.3 RMS results for function F_5

Importance level	Coefficient	p-value
ρ_1	−1.71	0.012
ρ_2	−2.08	0.010
ρ_3	−1.88	0.008
ρ_1^2	0.65	0.021
ρ_2^2	1.20	0.014
ρ_3^2	1.35	0.006
$\rho_1\rho_2$	0.83	0.018
$\rho_1\rho_3$	1.16	0.021
$\rho_2\rho_3$	1.18	0.011
K	−3.41	0.002

functions. Under such conditions, the final parameters are: $\hat{\rho}_1 = 0.85$, $\hat{\rho}_2 = 0.35$ and $\hat{\rho}_3 = 0.062$. Such optimal parameter values are kept for all tests throughout the chapter.

7.3 The Multi-modal States of Matter Search (MSMS)

In SMS, individuals emulate molecules which jointly interact among them by using evolutionary operators. The operations have been devised so that they simulate the phenomenon of the thermal-energy motion. The search strategy considers three stages: gas, liquid and solid. Each state represents a different exploration–exploitation association implemented by behavioral changes in the operators. As a result, SMS can substantially improve the balance between exploration–exploitation, yielding flexible search capabilities. In spite of such characteristics, the SMS method fails in providing multiple solutions within a single execution. In the MSMS approach, the original SMS is adapted to include multimodal capacities. Specifically, the approach is modified according to the following aspects: (1) the incorporation of a memory mechanism to efficiently register potential local optima; (2) the modification of the original SMS search strategy to increase the speed of detection for new local minima; (3) the inclusion of a depuration procedure at the end of each state to eliminate duplicated memory elements. Such adaptations are discussed below.

7.3.1 *Memory Mechanism*

In the MSMS evolution process, a population $\mathbf{P}^k$ ($\{\mathbf{p}_1^k, \mathbf{p}_2^k, \ldots, \mathbf{p}_N^k\}$) of N molecules (individuals) is operated from the initial point ($k = 1$) to a total *gen* number iterations ($k = gen$). Each molecule $\mathbf{p}_i^k$ ($i \in [1, \ldots, N]$) represents an n-dimensional vector $\left\{p_{i,1}^{k}, p_{i,2}^{k}, \ldots, p_{i,n}^{k}\right\}$ where each dimension corresponds to a decision variable of the optimization problem to be solved. The quality of each molecule $\mathbf{p}_i^k$ (candidate solution) is evaluated by an objective function $f\left(\mathbf{p}_i^k\right)$ whose final result represents the fitness value of $\mathbf{p}_i^k$. During the evolution process, MSMS maintains the best $\mathbf{p}^{best,k}$ and the worst $\mathbf{p}^{worst,k}$ molecules seen-so-far, such that

$$\begin{aligned} \mathbf{p}^{best,k} &= \underset{i\in\{1,2,\ldots,N\},\, a\in\{1,2,\ldots,k\}}{\arg\min} \left(f(\mathbf{p}_i^a)\right) \\ \mathbf{p}^{worst,k} &= \underset{i\in\{1,2,\ldots,N\},\, a\in\{1,2,\ldots,k\}}{\arg\max} \left(f(\mathbf{p}_i^a)\right). \end{aligned} \tag{7.9}$$

Global and local optima possess two significant properties: (1) they have a significant good fitness value and (2) they represent the best solutions within a certain area. Under these conditions, the memory mechanism allows efficiently registering potential global and local optima during the evolution process, involving a memory structure **M** and a storing process. **M** stores the potential global and local optima $\{\mathbf{m}_1, \mathbf{m}_2, \ldots, \mathbf{m}_T\}$ during the evolution process, being T the number of elements so-far that are contained in the memory **M**. On the other hand, the storage procedure indicates the rules that the molecules $\{\mathbf{p}_1^k, \mathbf{p}_2^k, \ldots, \mathbf{p}_N^k\}$ must fulfill in order to be captured as memory elements. The memory mechanism operates in two stages: initialization and capture.

7.3.1.1 Initialization Phase

This phase is applied only once within the optimization process. Such an operation is achieved in the first iteration ($k = 1$) of the evolution process. In the Initialization phase, the best molecule $\mathbf{p}_B$ of $\mathbf{P}^1$, in terms of its fitness value, is stored in the memory **M** ($\mathbf{m}_1 = \mathbf{p}_B$), where $\mathbf{p}_B = \underset{i\in\{1,2,\ldots,N\}}{\arg\min} \left(f(\mathbf{p}_i^1)\right)$.

7.3.1.2 Capture Phase

This phase is applied from the second iteration to the last iteration ($k = 2, 3, \ldots, gen$). At this stage, molecules $\{\mathbf{p}_1^k, \mathbf{p}_2^k, \ldots, \mathbf{p}_N^k\}$ corresponding to potential global and local optima are efficiently registered as memory elements $\{\mathbf{m}_1, \mathbf{m}_2, \ldots, \mathbf{m}_T\}$ according to their fitness quality and the distance to other promising solutions. In the operation, each molecule $\mathbf{p}_i^k$ of $\mathbf{P}^k$ is tested in order to evaluate if it must be captured as a memory element. The test considers two rules: (1) the significant fitness value rule and (2) the non-significant fitness value rule.

Significant fitness value rule

Under this rule, the quality of $\mathbf{p}_i^k$ is assessed regarding the worst member $\mathbf{m}^{worst}$ that is included in the memory $\mathbf{M}$, where $\mathbf{m}^{worst} = \underset{i \in \{1,2,\ldots,T\}}{\arg\max}\,(f(\mathbf{m}_i))$, for a minimization formulation. If the fitness value of $\mathbf{p}_i^k$ is better than $\mathbf{m}^{worst}(f(\mathbf{p}_i^k) < f(\mathbf{m}^{worst}))$, $\mathbf{p}_i^k$ is considered a potential global or local optima. The next step is to determine whether $\mathbf{p}_i^k$ represents a new optimum or it is very similar to an existent memory element $\{\mathbf{m}_1, \mathbf{m}_2, \ldots, \mathbf{m}_T\}$. Such a decision is specified by the evaluation of an acceptance probability function $\Pr(\delta_{i,u}, s)$ that depends, in one side, over the distance $\delta_{i,u}$ from $\mathbf{p}_i^k$ to the nearest memory element $\mathbf{m}_u$, and on the other side, over the current state s of the evolutionary process (Gas, liquid and solid). Under $\Pr(\delta_{i,u}, s)$, the probability that $\mathbf{p}_i^k$ would be part of $\mathbf{M}$ increases as the distance $\delta_{i,u}$ enlarges. Similarly, the probability that $\mathbf{p}_i^k$ would be analogous to an existent memory element $\{\mathbf{m}_1, \mathbf{m}_2, \ldots, \mathbf{m}_T\}$ increases as $\delta_{i,u}$ decreases. On the other hand, an indicator s that relates a numeric value with a state of matter ($s = 1$, gas, $s = 2$, liquid and $s = 3$, solid) is gradually modified during the algorithm to reduce the likelihood of permitting inferior solutions. The idea is that in the beginning of the evolutionary process (exploration) large distance differences can be considered while only small distance changes are accepted at the end of the optimization process.

In order to implement this process, the normalized distance $\delta_{i,q} (q \in [1, \ldots, T])$ is computed from $\mathbf{p}_i^k$ to all the elements of the memory $\mathbf{M}$ $\{\mathbf{m}_1, \mathbf{m}_2, \ldots, \mathbf{m}_T\}$. $\delta_{i,q}$ is calculated as follows:

$$\delta_{i,q} = \sqrt{\left(\frac{p_{i,1}^k - m_{q,1}}{b_1^{high} - b_1^{low}}\right)^2 + \left(\frac{p_{i,2}^k - m_{q,2}}{b_2^{high} - b_2^{low}}\right)^2 + \cdots + \left(\frac{p_{i,n}^k - m_{q,n}}{b_n^{high} - b_n^{low}}\right)^2}, \quad (7.10)$$

where $\{m_{q,1}, m_{q,2}, \ldots, m_{q,n}\}$ represent the n components of the memory element $\mathbf{m}_q$ whereas b_j^{high} and b_j^{low} indicate the low j parameter bound and the upper j parameter bound $(j \in [1, \ldots, n])$, respectively. One important property of the normalized distance $\delta_{i,q}$ is that its values fall into the interval [0,1].

By using the normalized distances $\delta_{i,q}$, the nearest memory element $\mathbf{m}_u$ to $\mathbf{p}_i^k$ is defined, with $\mathbf{m}_u = \underset{j \in \{1,2,\ldots,T\}}{\arg\min}\,(\delta_{i,j})$. Then, the acceptance probability function $\Pr(\delta_{i,u}, s)$ is calculated by using the following expression:

$$\Pr(\delta_{i,u}, s) = (\delta_{i,u})^s \quad (7.11)$$

In order to decide whether $\mathbf{p}_i^k$ corresponds to a new optimum or it is very similar to an existent memory element, a random number r_1 is produced inside the range [0,1] considering a uniform distribution. If r_1 is less than $\Pr(\delta_{i,u}, s)$, the molecule $\mathbf{p}_i^k$ is included in the memory $\mathbf{M}$ as a new optimum. Otherwise, it is considered that $\mathbf{p}_i^k$

is close to $\mathbf{m}_u$. Therefore, the memory $\mathbf{M}$ is updated by the competition between $\mathbf{p}_i^k$ and $\mathbf{m}_u$, according to their corresponding fitness values. Therefore, $\mathbf{p}_i^k$ would replace $\mathbf{m}_u$ in case of $f(\mathbf{p}_i^k) > f(\mathbf{m}_u)$. In addition, if $f(\mathbf{m}_u)$ is better than $f(\mathbf{p}_i^k)$, $\mathbf{m}_u$ remains with no change. The process of the significant fitness value rule is summarized by the following statement:

$$\mathbf{M} = \begin{cases} \mathbf{m}_{T+1} = \mathbf{p}_i^k & \text{with probability } \Pr(\delta_{i,u}, s) \\ \mathbf{m}_u = \mathbf{p}_i^k \text{ if } f(\mathbf{p}_i^k) < f(\mathbf{m}_u) & \text{with probability } 1 - \Pr(\delta_{i,u}, s) \end{cases} \quad (7.12)$$

To demonstrate the rule of the significant fitness value, Fig. 7.1 shows a simple minimization problem that involves a two-dimensional function $f(\mathbf{x})$ ($\mathbf{x} = \{x_1, x_2\}$). As an example, it is assumed a population $\mathbf{P}^k$ of two different particles ($\mathbf{p}_1^k$, $\mathbf{p}_2^k$), a memory with two memory elements ($\mathbf{m}_1$, $\mathbf{m}_2$) and the execution of the gas state (s = 1). According to Fig. 7.1, both particles $\mathbf{p}_1^k$ and $\mathbf{p}_2^k$ maintain a better fitness value than $\mathbf{m}_1$ which possesses the worst fitness value of the memory elements. Under such conditions, the rule has to be considered for both particles. In case of $\mathbf{p}_1^k$, the first step is to calculate the correspondent distances $\delta_{1,1}$ and $\delta_{1,2}$. $\mathbf{m}_1$ represents the nearest memory element to $\mathbf{p}_1^k$. Then, the acceptance probability function $\Pr(\delta_{1,1}, 1)$ is evaluated by using Eq. 7.13. Since the value of $\Pr(\delta_{1,1}, 1)$ has a high value, there exists a great probability that $\mathbf{p}_1^k$ becomes the next memory element ($\mathbf{m}_3 = \mathbf{p}_1^k$). On the other hand, for $\mathbf{p}_2^k$, $\mathbf{m}_2$ corresponds to the nearest memory element. As $\Pr(\delta_{2,2}, 1)$ is very low, there exists a great probability that $\mathbf{p}_2^k$ competes with $\mathbf{m}_2$ for a place within $\mathbf{M}$. In such a case, $\mathbf{m}_2$ remains fixed considering that $f(\mathbf{m}_2) < f(\mathbf{p}_2^k)$.

Non-significant fitness value rule

Under the new rule, local optima with low fitness values are detected. It operates when $f(\mathbf{p}_i^k) \geq f(\mathbf{m}^{worst})$. First, to test the particles that could represent local optima and which must be ignored due to their very low fitness value. Then, if the particle represent a possible local optimum, its inclusion inside the memory $\mathbf{M}$ is explored.

The decision on whether $\mathbf{p}_i^k$ corresponds to a local optimum or not is determined by a probability function E which is based on the relationship between $f(\mathbf{p}_i^k)$ and the so-far valid fitness value interval $(f(\mathbf{p}^{worst,k}) - f(\mathbf{p}^{best,k}))$. Therefore, the probability function E is defined as follows:

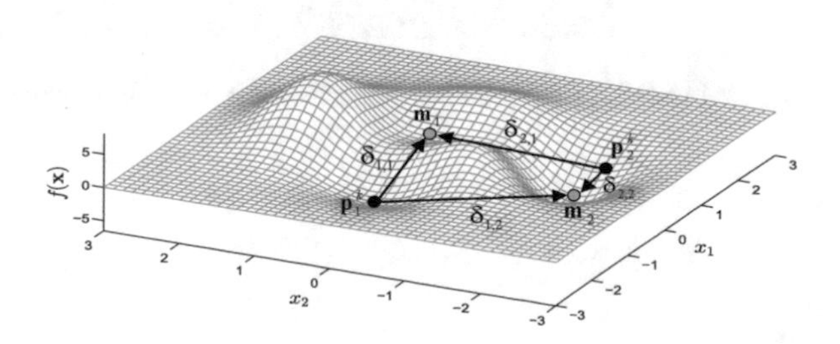

Fig. 7.1 Graphical illustration of the significant fitness value rule process

$$v(\mathbf{p}_i^k, \mathbf{p}^{best,k}, \mathbf{p}^{worst,k}) = 1 - \frac{f(\mathbf{p}_i^k) - f(\mathbf{p}^{best,k})}{f(\mathbf{p}^{worst,k}) - f(\mathbf{p}^{best,k})},$$
$$E(v) = \begin{cases} v & 0.5 \leq v \leq 1 \\ 0 & 0 \leq v < 0.5 \end{cases}, \tag{7.13}$$

where $\mathbf{p}^{best,k}$ and $\mathbf{p}^{worst,k}$ represent the best and worst molecule seen-so-far, respectively. To decide the category of $\mathbf{p}_i^k$, a uniform random number r_2 is generated inside the interval [0,1]. If r_2 is less than E, the molecule $\mathbf{p}_i^k$ is considered as a new local optimum. Otherwise, it must be ignored. Under E, the so-far valid fitness value interval $(f(\mathbf{p}^{worst,k}) - f(\mathbf{p}^{best,k}))$ is divided into two sections: I and II (see Fig. 7.2). Considering this division, the function E assigns a valid probability (greater than zero) only to those molecules that fall into the zone of the best individuals (part I) in terms of their fitness value. Such a probability value increases as the quality of the fitness value enlarges. The complete procedure can be reviewed in the Algorithm 1.

If the particle represent a possible local optimum, its inclusion inside the memory $\mathbf{M}$ is explored. To decide if $\mathbf{p}_i^k$ could correspond to a new memory element, another procedure that is similar to the significant fitness value rule process is considered. Therefore, it is calculated the normalized distance $\delta_{i,q}(q \in [1, \ldots, T])$ from $\mathbf{p}_i^k$ to all the elements of the memory $\mathbf{M}$ $\{\mathbf{m}_1, \mathbf{m}_2, \ldots, \mathbf{m}_T\}$, according to Eq. 7.12. Afterwards, the nearest distance $\delta_{i,u}$ to $\mathbf{p}_i^k$ is determined. Then, by using $\Pr(\delta_{i,u}, s)$ (Eq. 7.13), the following rule can be thus applied:

$$\mathbf{M} = \begin{cases} \mathbf{m}_{T+1} = \mathbf{p}_i^k & \text{with probability } \Pr(\delta_{i,u}, s) \\ \text{no change} & \text{with probability } 1 - \Pr(\delta_{i,u}, s) \end{cases} \tag{7.14}$$

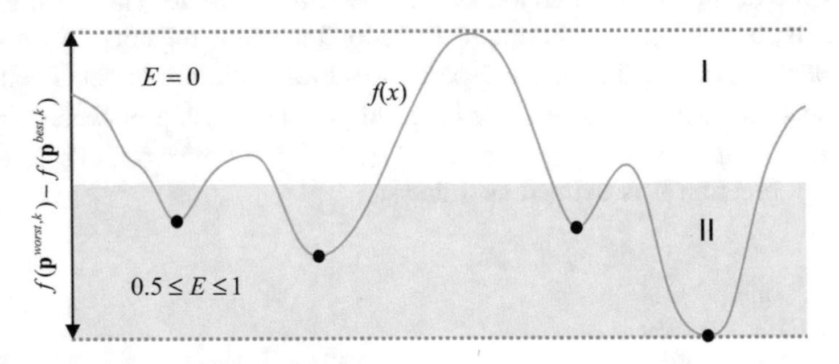

Fig. 7.2 Effect of the probability function E in a simple example

Under this rule, a uniform random number r_3 is generated within the range [0, 1]. If r_3 is less than $\Pr(\delta_{i,u}, s)$, the molecule $\mathbf{p}_i^k$ is included in the memory $\mathbf{M}$ as a new optimum. Otherwise, the memory does not change.

Algorithm 1. Non-significant fitness value rule procedure		
1:	**Input:** $\mathbf{p}_i^k, \mathbf{p}^{best,k}, \mathbf{p}^{worst,k}$	
2:	Calculate $v(\mathbf{p}_i^k, \mathbf{p}^{best,k}, \mathbf{p}^{worst,k}) = 1 - \frac{f(\mathbf{p}_i^k)-f(\mathbf{p}^{best,k})}{f(\mathbf{p}^{worst,k})-f(\mathbf{p}^{best,k})}$	
3:	Calculate $E(v) = \begin{cases} v & 0.5 \leq v \leq 1 \\ 0 & 0 \leq v < 0.5 \end{cases}$	
5:	**if** (rand(0,1)<=E) **then**	
6:	$\mathbf{p}_i^k$is considered a local optimum	With probability E
7:	**else**	
8:	$\mathbf{p}_i^k$is ignored	With probability $1 - E$
9:	**end if**	

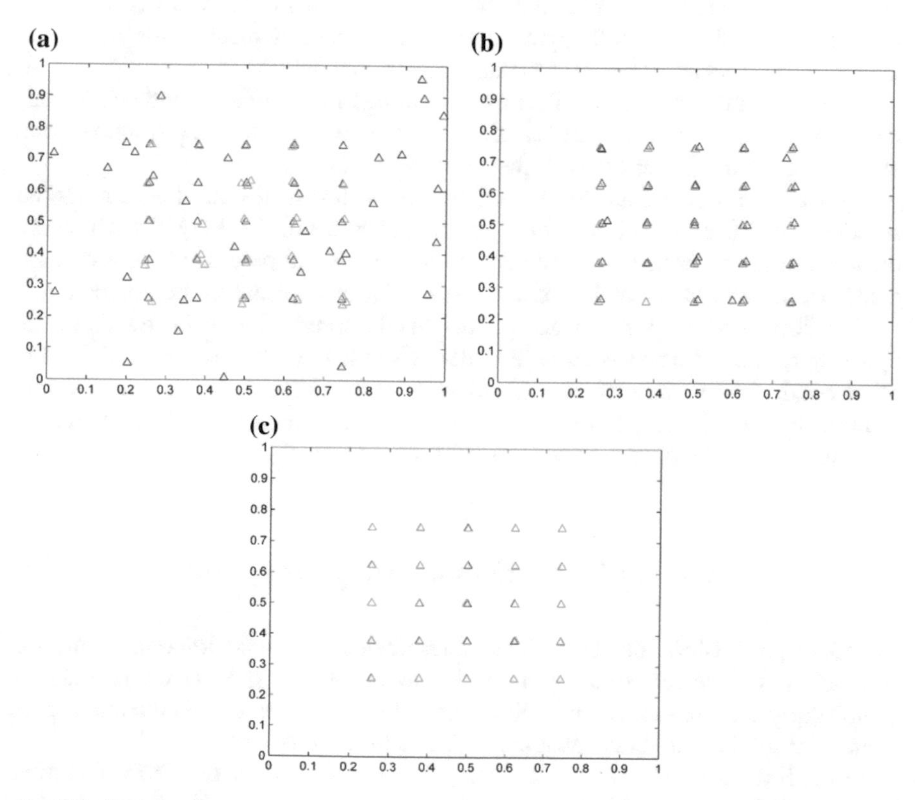

Fig. 7.3 Optimization process. **a** Gas state, **b** liquid state and **c** solid state. Optimal positions, molecules (search positions) and memory elements are represented by red, blue and green triangles, respectively

In order to illustrate the performance of memory **M**, Fig. 7.3 shows the optimization process, where the elements (candidate solutions) that are contained in the memory can be seen evolving through the three different states: gas, liquid and solid. In the figure, a simple multimodal optimization problem is solved, where the optimal positions, molecules (search positions) and memory elements are represented by red, blue and green triangles, respectively.

7.3.1.3 Memory Mechanism and Its Similarity with Other EA Methods

The use of a memory mechanism in a stochastic algorithm has been already reported in the literature. One example is the Tabu Search (TS) method [72, 73]. It is a local search algorithm that allows non-improving solutions when a local optimum is encountered, in the hope that in this way the solution will be improved. An important component of TS is the memory known as Tabu list which stores the solutions explored throughout the search process. This information is used during the search strategy to avoid the selection of previously visited solutions. Although TS and the MSMS approach consider a memory mechanism to conduct their search strategies, there exist considerable differences:

TS store solutions located only in a determined neighborhood without considering their fitness values whereas the MSMS algorithm register solutions that exhibit a good quality in terms of their fitness values.

TS uses the stored solutions to prevent the return to the most recent visited solutions in order to avoid cycling. On the other hand, the MSMS multimodal method considers the memory elements to accelerate the process of local optimal detection through the attraction of candidate solutions present in the population.

The Tabu list is updated when a new neighboring solution is found. In the updating process, the new solution is added whereas the oldest is removed. On the other hand, the memory of the MSMS method is updated always when a new solution that exhibits a good quality is presented. In contrast to TS, the updating process does not involve the elimination of any element.

7.3.2 Modification of the Original SMS Search Strategy

In the original SMS, the best element represents the most important part for conducting the search strategy. In order to accelerate the search process of promising local minima, in our method, the optimization strategy is modified to be conducted by the individuals that are included in the memory.

In the SMS method, the search strategy is mainly ruled by the vector direction operator. Under this operator, for each n-dimensional molecule $\mathbf{p}_i^k$ from the population $\mathbf{P}^k$, it is assigned an n-dimensional direction vector $\mathbf{d}_i^k$ ($\mathbf{D}^k = \{\mathbf{d}_1^k, \mathbf{d}_2^k, \ldots, \mathbf{d}_{N_p}^k\}$)

which stores the vector that controls the particle movement. Initially, all the direction vectors ($\mathbf{D}^1 = \{\mathbf{d}_1^1, \mathbf{d}_2^1, \ldots, \mathbf{d}_{N_p}^1\}$) are randomly chosen within the range of [−1,1]. The new direction vector $\mathbf{d}_i^{k+1}$ for each molecule $\mathbf{p}_i^k$ is iteratively computed (Eq. 7.1) through a combination between the current direction vector $\mathbf{d}_i^k$ and an attraction vector $\mathbf{a}_i$. In the original SMS method, $\mathbf{a}_i$ is calculated as follows:

$$\mathbf{a}_i = \frac{(\mathbf{p}^{best,k} - \mathbf{p}_i^k)}{\|\mathbf{p}^{best,k} - \mathbf{p}_i^k\|} \tag{7.15}$$

By using $\mathbf{a}_i$, each particle $\mathbf{p}_i^k$ is moved towards the best individual $\mathbf{p}^{best,k}$ seen so-far. This effect allows incorporating interesting convergence properties to SMS when the trial considers only one optimum. Nonetheless, this operation cannot be appropriate for multiple optimum localization. To accelerate the detection of promising local minima in our method, the attraction vector $\mathbf{a}_i$ is modified to be influenced by the members included in the memory $\mathbf{M}$.

Under the new $\mathbf{a}_i$, each particle $\mathbf{p}_i^k$ is moved towards the nearest memory element $\mathbf{m}_u$ to $\mathbf{p}_i^k$. Therefore, the new attraction vector $\mathbf{a}_i^{new}$ is redefined as follows:

$$\mathbf{a}_i^{new} = \frac{(\mathbf{m}_u - \mathbf{p}_i^k)}{\|\mathbf{m}_u - \mathbf{p}_i^k\|} \tag{7.16}$$

Figure 7.4 illustrates the differences between the original SMS method and the modified version for the minimization of a two-dimensional function $f(\mathbf{x})$ ($\mathbf{x} = \{x_1, x_2\}$), assuming a population $\mathbf{P}^k$ of five different particles (white points). Figure 7.4a shows the attraction vectors generated by the original SMS method among molecules (white points) and the best molecule seen so-far (blue point). On the other hand, Fig. 7.4b exhibits the attraction vectors produced by the MSMS approach. According to such vectors, the current molecules of $\mathbf{P}^k$ are attracted to the

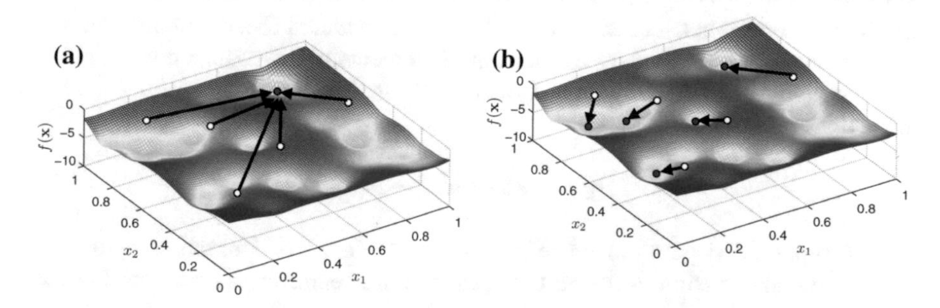

Fig. 7.4 Search strategy differences: **a** the original SMS and **b** the MSMS approach

nearest memory elements (red points) that are contained in **M**. This behavior supports not only an effective multiple-optima encompassing but also a faster computation.

7.3.3 Depuration Procedure

During the evolution process, the memory **M** includes multiple individuals (molecules). Since such elements could represent the same local optimum, a depuration procedure is incorporated at the end of each state to eliminate similar memory elements. The inclusion of this procedure allows (a) reducing the computational overhead during each state and (b) improving the search strategy by maintaining only important memory elements.

Memory elements tend to concentrate around optimal points (good fitness values) whereas element concentrations are enclosed by areas holding bad fitness values. The main idea in the depuration process is to determine the distances among concentrations. Such distances, considered as depuration ratios, are later employed to delete all elements inside them, except for the best element regarding their fitness values.

The method used by the depuration procedure to determine the distance between two concentrations is based on the element comparison. Under this process, the concentration that corresponds to the best element and the conglomeration of the nearest optimum in the memory are compared. In the process, the element $\mathbf{m}^{best}$ is contrasted with the memory member $\mathbf{m}_b$ that belongs to one of both concentrations (where $\mathbf{m}^{best} = \arg\min_{i \in \{1,2,\ldots,T\}} (f(\mathbf{m}_i))$). If the fitness value of the middle point $f((\mathbf{m}^{best} + \mathbf{m}_b)/2)$ between both members is not as good as $f(\mathbf{m}^{best})$ and $f(\mathbf{m}_b)$, then the element $\mathbf{m}_b$ is part of the $\mathbf{m}^{best}$ concentration. But, if $f((\mathbf{m}^{best} + \mathbf{m}_b)/2)$ is of lower quality than both, the element $\mathbf{m}_b$ is considered part of the neighbor concentration. Therefore, if $\mathbf{m}_b$ and $\mathbf{m}^{best}$ belong to different conglomerators, the Euclidian distance between $\mathbf{m}_b$ and $\mathbf{m}^{best}$ can be considered as a depuration ratio. In order to avoid the involuntary elimination of elements in the nearest concentration, the depuration ratio D_R is lightly shortened. Thus, the depuration ratio r is characterized as follows:

$$D_R = 0.85 \cdot \left\| \mathbf{m}^{best} - \mathbf{m}_b \right\|, \tag{7.17}$$

The proposed depuration procedure only considers the depuration ratio r between the concentration of the best element and the nearest concentration. In order to determine all ratios, pre-processing and post-processing methods must be incorporated and iteratively executed.

The pre-processing method, must (1) obtain the best element $\mathbf{m}^{best}$ from the memory in terms of its fitness value, (2) calculate the distances from $\mathbf{m}^{best}$ to the other memory elements and (3) arrange the distances regarding their magnitude. This set of tasks allows identifying both concentrations: the one belonging to the best element and the one belonging to the nearest optimum. These operations must be completed before the calculation of the depuration ratio D_R. Such concentrations are characterized by the members with the shortest distances to $\mathbf{m}^{best}$. Once D_R has been calculated, it is important to delete all the elements belonging to the concentration of the best element. This task is executed as a post-processing method to configure the memory for the next step. Therefore, the complete depuration procedure can be considered as an repetitive process that at each step determines the distance of the concentration of $\mathbf{m}^{best}$ to the concentration of the nearest optimum.

An especial case can be considered when only one concentration is contained within the memory. This case can happen because the optimization formulation presents a single optimum or because all the other concentrations have been already detected. Under such circumstances, the condition, where $f(\mathbf{m}^{best})$ and $f(\mathbf{m}_b)$ are better than $f((\mathbf{m}^{best}+\mathbf{m}_b)/2)$, would be never satisfied.

In order to find the distances among concentrations, the depuration procedure conducts the following procedure:

1. Define two new temporal vectors $\mathbf{Z}$ and $\mathbf{Y}$. The vector $\mathbf{Z}$ will hold the results of the iterative operations whereas $\mathbf{Y}$ will contain the final memory configuration. The vector $\mathbf{Z}$ is initialized with the elements of $\mathbf{M}$ that have been sorted according to their fitness values, so that the top element corresponds to the best one. $\mathbf{Y}$ is initialized empty.
2. Store the best element $\mathbf{z}_1$ of the current $\mathbf{Z}$ in $\mathbf{Y}$.
3. Calculate the Euclidian distances $\Delta_{1,j}$ between $\mathbf{z}_1$ and the rest of elements from $\mathbf{Z}$ ($j \in \{2,\ldots,|\mathbf{Z}|\}$), where $|\mathbf{Z}|$ represents the number of elements in $\mathbf{Z}$.
4. Sort the distances $\Delta_{1,j}$ according to their magnitude. Therefore, a new index a is incorporated to each distance $\Delta_{1,j}^{a}$, where a indicate the place of $\Delta_{1,j}$ after the sorting operation ($a = 1$ represents the shortest distance).
5. Calculate the depuration ratio r:

for $q = 1$ **to** $|\mathbf{Z}| - 1$
Obtain the element $\mathbf{z}_j$ corresponding to the distance $\Delta_{1,j}^{q}$
Compute $f((\mathbf{z}_1+\mathbf{z}_j)/2)$
  if $(f((\mathbf{z}_1+\mathbf{z}_j)/2) > f(\mathbf{z}_1)$ and $f((\mathbf{z}_1+\mathbf{z}_j)/2) > f(\mathbf{z}_j))$
  $D_R = 0.85 \cdot \left\|\mathbf{x}_1 - \mathbf{x}_j\right\|$
  break
  end if
    if $q = |\mathbf{Z}| - 1$
    There is only one concentration
    end if
end for

6. Remove all elements inside D_R from $\mathbf{Z}$.

7. Sort the elements of $\mathbf{Z}$ according to their fitness values.
8. Stop, if there are even more data conglomerations, otherwise return to step 2.

At the end of the above process, the vector $\mathbf{Y}$ will contain the depurated memory which would be used in the next state of matter computation or as a final result of the multi-modal problem. In order to illustrate the depuration procedure, Fig. 7.5 shows a simple minimization problem that involves two different optimal points (concentrations). As an example, it is assumed a memory $\mathbf{M}$ with six memory elements whose positions are shown in Fig. 7.5a. According to the depuration procedure, the first stage is (1) to build the vector $\mathbf{Z}$ and (2) to calculate the corresponding distances $\Delta^{a}_{1,j}$ among the elements. Following such operation, the vector $\mathbf{Z}$ and the set of distances are configured as $\mathbf{Z} = \{\mathbf{m}_5, \mathbf{m}_1, \mathbf{m}_3, \mathbf{m}_4, \mathbf{m}_6, \mathbf{m}_2\}$ and $\left\{\Delta^{1}_{1,2}, \Delta^{2}_{1,3}, \Delta^{3}_{1,5}, \Delta^{4}_{1,4}, \Delta^{5}_{1,6}\right\}$, respectively. Figure 7.5b shows the configuration of $\mathbf{X}$ whereas, for sake of easiness, only two distances $\Delta^{1}_{1,2}$, and $\Delta^{3}_{1,5}$ have been represented. Then, the depuration ratio R is calculated. This method is an iterative process that begins with the shortest distance $\Delta^{1}_{1,2}$. The distance $\Delta^{1}_{1,2}$ (see Fig. 7.5c), corresponding to $\mathbf{z}_1$ and $\mathbf{z}_2$, produces the evaluation of their medium point u $((\mathbf{z}_1 + \mathbf{z}_2)/2)$. Since $(f(u) < f(\mathbf{z}_1))$ but $(f(u) > f(\mathbf{z}_2))$, the point $\mathbf{z}_2$ is contained in the same concentration as $\mathbf{z}_1$. We obtain the same conclusion for $\Delta^{2}_{1,3}$ in case of $\mathbf{z}_3$, after observing the point v. For $\Delta^{3}_{1,5}$, the point

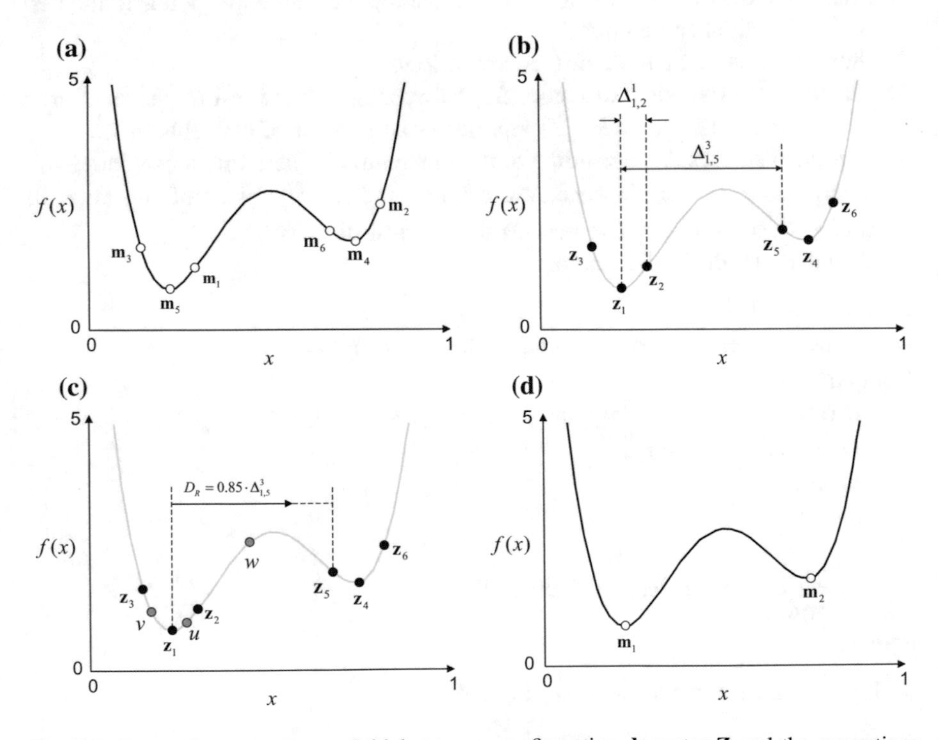

Fig. 7.5 Depuration procedure. **a** Initial memory configuration, **b** vector $\mathbf{Z}$ and the separations $\Delta^{a}_{1,j}$, **c** the calculation of the depuration ratio D_R and **d** the final memory configuration

w is produced. Since $f(w)$ is worse than $f(\mathbf{z}_1)$ and $f(\mathbf{z}_5)$, the point $\mathbf{z}_5$ is considered as part of the concentration corresponding to the next optimum. The iterative process ends here, after considering that the same result is produced with $\Delta_{1,4}^{4}$ and $\Delta_{1,6}^{5}$, for $\mathbf{z}_4$ and for $\mathbf{z}_6$ respectively. Therefore, the depuration ratio D_R is calculated as the 85% of the distance $\Delta_{1,5}^{3}$. Once the elements inside of D_R have been removed from $\mathbf{Z}$, the same process is applied to the new $\mathbf{Z}$. As a result, the final state of the memory is shown in Fig. 7.5d.

7.3.4 Discussion About the MSMS Algorithm

In the MSMS approach, the original SMS is adapted to include multimodal capacities. Specifically, SMS is modified including the following aspects: (1) the incorporation of a memory mechanism to efficiently register potential local optima; (2) the modification of the original SMS search strategy to increase the speed of detection for new local minima; (3) the inclusion of a depuration procedure at the end of each state to eliminate duplicated memory elements.

From the three new incorporations, the inclusion of the memory and the depuration process represent the systems that provide the multimodal capacities to the final MSMS approach. On the other hand, the modification of the original SMS search strategy permits only to accelerate the process of detecting new potential solutions (global or local optima). Hence, the memory and its depuration could be considered as the most significant adaptations for including multimodal properties to a standard EA method.

In MSMS, the final configuration of the memory (its elements) represents the solution to the multimodal optimization problem. For this reason, the memory must contain only those solutions considered as the best representative global and local optima. However, during the search process, the memory stores several individuals which could represent the same local optimum. Therefore, a depuration procedure is necessary to eliminate similar memory elements. The depuration process is a generic procedure and must be applied in order to deliver a reliable set of solutions for a multimodal optimization problem. Without its incorporation, the memory would contain a very big set of elements where most of them represents the same position. Furthermore, under such conditions, the algorithm would increase its computational overhead as a consequence of the exaggerated number of elements considered in the computations.

7.4 Experimental Results

An illustrative set of 18 functions have been used to examine the performance of our approach. They are divided in two sets: Fixed Functions and composition functions. Fixed functions represent optimization problems formulated in two dimensions. On the other hand, composite function are multidimensional problems which are constructed as a weighted aggregation of basic functions. These functions have been recently used to prove the performance of multimodal algorithms in the literature [21]. The set of functions used in the experimental study are shown in Tables 7.10 and 7.11 in the Appendix. Table 7.10 shows the fixed functions which involve problems from f_1 to f_{14} while Table 7.11 defines the composition functions that corresponds to problems F_1 to F_4. In Tables 7.10 and 7.11, n represents the dimension in which the function is operated, **NO** characterizes the number of optima and S is the defined search space.

7.4.1 Experimental Methodology

In the study, five performance elements are compared: the effective peak number (*EPN*), the maximum peak ratio (*MPR*), the distance accuracy (*DA*), the peak accuracy (*PA*) and the computational time (*CT*). The first four indexes assess the accuracy of the solution whereas the latter measures the computational effort.

The effective peak number (*EPN*) defines the number of detected peaks. An optimum $\mathbf{o}_j$ is considered as detected if the distance between the identified solution $\mathbf{z}_j$ and the optimum $\mathbf{o}_j$ is less than 0.01 ($\left\|\mathbf{o}_j - \mathbf{z}_j\right\| < 0.01$). The maximum peak ratio (*MPR*) is employed to evaluate the quality and the number of identified optima. It is defined as follows:

$$MPR = \frac{\sum_{i=1}^{t} f(\mathbf{z}_i)}{\sum_{j=1}^{q} f(\mathbf{o}_j)}, \tag{7.18}$$

where t represents the number of detected solutions (identified optima) for the algorithm under testing and q the number of true optima contained in the function. The peak accuracy (*PA*) specifies the total error produced between the identified solutions and the true optima. Therefore, *PA* is calculated as follows:

$$PA = \sum_{j=1}^{q} \left| f(\mathbf{o}_j) - f(\mathbf{z}_j) \right| \tag{7.19}$$

Peak accuracy (*PA*) may lead to incorrect results, mainly if the peaks are close to each other or hold an identical height. To eliminate this inconvenience, the distance accuracy (*DA*) is employed. *DA* is computed as *PA*, but fitness values are interchanged by the Euclidian distance. *DA* is thus calculated by the following formulation:

$$DA = \sum_{j=1}^{q} \left\| \mathbf{o}_j - \mathbf{z}_j \right\| \qquad (7.20)$$

In order to locate several optima, multimodal optimization algorithms attempt to conserve the individual diversity. Additional to the other indexes, in this work, the formulation proposed in [47] has been used to assess the population diversity provided by each algorithm. Therefore, the diversity at the *k* iteration is calculated as follows:

$$\text{Diversity}_k = \frac{\sum_{i=1}^{N-1} \sum_{j=i+1}^{N} dis(\mathbf{p}_i^k, \mathbf{p}_j^k)}{N \cdot (N-1)/2}, \qquad (7.21)$$

where N is the population size, *dist* represents the Euclidian distance, and $\mathbf{p}_i^k, \mathbf{p}_j^k$ are solutions in the population.

The experiments compare the performance of MSMS against the Crowding Differential Evolution (CDE) [29], the Fitness Sharing Differential Evolution (SDE) [28, 29], the Clearing Procedure (CP) [71], the Elitist-population strategy (AEGA) [36], the Clonal Selection algorithm (CSA) [42], the artificial immune network (AiNet) [43], the Multimodal Gravitational Search algorithm (MGSA) [31] and the Region-Based Memetic method (RM) [47].

Since the approach solves real-valued multimodal problems and a fair comparison must be assured, we have used for the GA-approaches a consistent real coding variable representation and uniform operators for crossover and mutation. The crossover probability $P_c = 0.8$ and the mutation probability $P_m = 0.1$ have been used. We have employed the standard tournament selection operation with a tournament size = 2 for implementing the Sequential Fitness Sharing, the CP and the AEGA strategy. The values of the parameters for the AiNet algorithm have been defined as it is suggested in [43], with the mutation strength $\beta = 100$, the suppression threshold $\sigma_{s(aiNet)} = 0.2$ and the update rate $d = 40\%$. Algorithms based on DE use a scaling element $F = 0.5$ and a crossover probability $p_c = 0.9$. The crowding-DE employs a crowding factor $CF = 50$ and the sharing-DE considers $\alpha = 1.0$ with a share radius $\sigma_{share} = 0.1$. It is important to point out that the configuration of each method is set according to its reported guidelines. All these settings represent the best possible performance of the algorithms according to their own reported references.

In the case of the MSMS algorithm, the parameters are set to $\rho = [0.85, 0.35, 0.062]$, $\beta = [0.8, 0.4, 0.1]$, $\alpha = [0.8, 0.2, 0]$ and $H = [0.9, 0.2, 0]$, where the first element of each vector corresponds to the gas state configuration, the second element to the liquid state and the third element to the solid state. Once they have been all experimentally determined, they are kept for all the test functions through all experiments.

To avoid relating the optimization outcomes to the selection of a particular initial population and to conduct fair conclusions, we perform 50 execution for each test, starting from several randomly selected points in the search space.

The experimental section have been divided into two groups. The first one considers the fixed functions, while the second gathers the composition functions.

7.4.2 Comparing MSMS Performance for Fixed Functions (Low Dimensional)

This section presents the performance comparison for different algorithms solving the multimodal problems f_1–f_{14} that are shown in Table 7.10. The aim is to determine whether MSMS is more efficient and effective than other existing algorithms for finding all multiple optima. All the algorithms employ a population of 50 elements by using 2.5×10^4 function evaluations. This stop criterion has been decided to keep compatibility with similar works published in the literature.

For the sake of clarity, Tables 7.4 and 7.5 present the summarized performance results among the algorithms, for functions f_1–f_7 and f_8–f_{14}, respectively. The tables report the comparison in terms of the effective peak number (*EPN*), the maximum peak ratio (*MPR*), the peak accuracy (*PA*) and the distance accuracy (*DA*). The results are analyzed by considering 50 different executions.

From Table 7.4, according to the *EPN* index, MSMS always finds better or equally optimal solutions for the multimodal problems f_1–f_4. In case of function f_1, the CDE, RM and the MSMS algorithms can find all optima of f_1. For function f_2, only CSA, AEGA and AiNet have not been able to detect all the optima values each time. For function f_3, only MSMS can get all optima at each run. In case of function f_4, most of the algorithms cannot get any better results but MSMS which can yet find most of the optima. For function f_5, CDE, CP, CSA and AiNet maintain a similar performance whereas SDE, AEGA, MGSA, RM and MSMS possess the best *EPN* values. In case of f_6, almost all methods present a similar performance; however, only the CDE, CP, MGSA, RM and MSMS algorithms have been able to detect all optima. For function f_7, the MSMS algorithm detects most of the optima whereas the rest of the methods reach different performance levels. By analyzing the *MPR* index in Table 7.4, MSMS has obtained the best score for all the multimodal problems. On the other hand, the other approaches present different accuracies, with MGSA and RM being the most consistent. In case of the *PA* index, MSMS presents the best performance. Since the *PA* index evaluates the accumulative differences of fitness values, it could drastically change when one or several peaks are not identified (function f_3) or when the function under testing presents peaks with high values (function f_4). For the case of *DA*, it is clear that the MSMS algorithm presents the best performance obtaining the shortest distances among the detected optima. It can be easily deduced from such results that the MSMS algorithm is able to produce better search locations (i.e. a better trade-off between

Table 7.4 Results of functions f_1–f_7 of Table 7.10 with $n = 2$

Function	Algorithm	*EPN*	*MPR*	*PA*	*DA*
f_1	CDE	3 (0)	0.9545 (0.0011)	0.1124 (0.0224)	0.1571 (0.0321)
	SDE	2.2 (0.51)	0.9001 (0.0842)	1.6492 (1.1231)	0.2341 (0.0851)
	CP	2.0 (0.10)	0.8241 (0.1011)	1.8974 (1.2576)	0.3871 (0.0891)
	AEGA	2.8 (0.2)	0.9487 (0.0088)	0.1208 (0.4157)	0.2007 (0.0101)
	CSA	2.90 (0.18)	0.9111 (0.0470)	1.4308 (1.0287)	0.2025 (0.0062)
	AiNet	2.95 (0.15)	0.9051 (0.0885)	1.3685 (1.0118)	0.1802 (0.0097)
	MGSA	2.85 (0.2)	0.9187 (0.0047)	0.9987 (1.1287)	0.2187 (0.0259)
	RM	3 (0)	0.9687 (0.0982)	0.2157 (0.0322)	0.1118 (0.0913)
	MSMS	3 (0)	1 (0)	0.0107 (0.0067)	0.0874 (0.0100)
f_2	CDE	12 (0)	1 (0)	0.0227 (0.0217)	0.3174 (0.0667)
	SDE	12 (0)	1 (0)	0.0314 (0.0114)	0.4014 (0.0917)
	CP	12 (0)	1 (0)	0.0512 (0.0081)	0.3041 (0.0758)
	AEGA	11.8 (0.2)	0.9557 (0.0217)	0.1121 (0.0107)	0.4820 (0.0327)
	CSA	11.90 (0.12)	0.9214 (0.0201)	0.1298 (0.0461)	0.4541 (0.0308)
	AiNet	11.92 (0.21)	0.9311 (0.0101)	0.1002 (0.0360)	0.3961 (0.0400)
	MSGA	12 (0)	1 (0)	0.0687 (0.0604)	0.3108 (0.0924)
	RM	12 (0)	1 (0)	0.0704 (0.0491)	0.2974 (0.0501)
	MSMS	12 (0)	1 (0)	0.0124 (0.0094)	0.1174 (0.0148)
f_3	CDE	22.07 (2.11)	0.8421 (0.0741)	143.27 (101.41)	10.657 (5.9841)
	SDE	18.82 (1.10)	0.6687 (0.0540)	177.32 (98.32)	15.021 (2.4170)
	CP	19.51 (2.52)	0.7008 (0.0981)	160.12 (61.07)	13.216 (1.1434)
	AEGA	18.04 (3.51)	0.6271 (0.0740)	180.74 (71.52)	16.010 (1.8791)
	CSA	16.80 (1.85)	0.4967 (0.0927)	195.21 (100.2)	18.527 (5.8720)
	AiNet	17.66 (2.87)	0.5274 (0.0811)	190.07 (84.52)	17.217 (3.8541)
	MSGA	22.74 (2.35)	0.8800 (0.0141)	137.87 (40.41)	9.9140 (2.004)
	RM	23.22 (1.27)	0.8841 (0.0124)	120.14 (20.70)	6.0047 (1.0214)
	MSMS	25 (0)	1 (0)	10.11 (5.87)	2.0874 (1.4071)
f_4	CDE	3.21 (1.20)	0.4824 (0.1017)	384.10 (154.12)	207.074 (81.11)
	SDE	3.41 (1.54)	0.5281 (0.1195)	311.17 (171.01)	197.271 (77.21)
	CP	3.04 (2.01)	0.4141 (0.0547)	401.27 (95.14)	225.874 (54.04)
	AEGA	3.50 (1.18)	0.6004 (0.0967)	195.54 (81.07)	147.574 (23.41)
	CSA	3.00 (0.50)	0.4001 (0.0274)	421.07 (100.87)	237.88 (52.07)
	AiNet	3.20 (1.05)	0.4800 (0.0195)	391.74 (97.02)	219.42 (36.87)
	MSGA	5.04 (1.41)	0.8007 (0.0274)	127.84 (57.04)	100.51 (25.11)
	RM	5.88 (1.81)	0.8997 (0.0576)	80.41 (10.87)	79.97 (9.87)
	MSMS	6.71 (0.30)	0.9524 (0.0410)	18.98 (9.97)	21.85 (7.96)
f_5	CDE	21.07 (3.54)	0.4725 (0.0181)	1.1074 (0.0280)	15.21 (2.007)
	SDE	30.87 (1.84)	0.6598 (0.0307)	0.6897 (0.0114)	4.071 (0.8741)
	CP	20.90 (1.01)	0.4457 (0.0100)	1.3985 (0.0214)	16.00 (1.087)
	AEGA	29.04 (2.84)	0.6400 (0.0201)	0.671 (0.0174)	4.121 (0.984)

(continued)

Table 7.4 (continued)

Function	Algorithm	*EPN*	*MPR*	*PA*	*DA*
	CSA	22.99 (3.14)	0.4951 (0.0257)	0.9841 (0.0274)	13.231 (2.874)
	AiNet	26.21 (1.06)	0.6287 (0.0291)	0.7181 (0.0171)	6.987 (1.820)
	MSGA	31.21 (2.11)	0.7932 (0.0361)	0.6001 (0.0218)	3.887 (0.847)
	RM	33.00 (0.50)	0.9101 (0.0521)	0.3002 (0.0101)	3.047 (0.102)
	MSMS	35.66 (1.04)	0.9821 (0.0104)	0.2201 (0.0431)	3.005 (0.137)
f_6	CDE	6 (0)	0.9612 (0.0221)	0.082 (0.0091)	0.1291 (0.0173)
	SDE	5.57 (0.18)	0.9099 (0.0247)	0.092 (0.0042)	0.1342 (0.0194)
	CP	6 (0)	0.9589 (0.0221)	0.088 (0.0025)	0.1274 (0.0230)
	AEGA	5.0 (0.50)	0.8801 (0.0185)	0.1002 (0.0562)	0.2008 (0.0321)
	CSA	4.20 (0.34)	0.8006 (0.0223)	0.2011 (0.0130)	0.2884 (0.0119)
	AiNet	5.10 (0.8)	0.8902 (0.0301)	0.0997 (0.0092)	0.1892 (0.0072)
	MSGA	6 (0)	0.9906 (0.0129)	0.053 (0.0020)	0.0251 (0.0011)
	RM	6 (0)	0.9950 (0.0093)	0.0040 (0.0018)	0.0180 (0.0024)
	MSMS	6 (0)	0.9971 (0.0051)	0.0029 (0.0034)	0.0151 (0.0031)
f_7	CDE	28.01 (2.41)	0.6076 (0.0131)	2.1007 (0.2311)	338.77 (23.81)
	SDE	33.22 (2.65)	0.7001 (0.0248)	1.8842 (0.0862)	247.21 (32.09)
	CP	33.50 (3.40)	0.7098 (0.0157)	1.7801 (0.0286)	220.97 (35.22)
	AEGA	30.83 (1.90)	0.6797 (0.0156)	2.0121 (0.0430)	290.57 (21.08)
	CSA	31.70 (2.20)	0.6811 (0.0500)	1.9107 (0.0742)	280.11 (30.87)
	AiNet	34.50 (2.01)	0.7287 (0.0240)	1.5089 (0.0331)	200.81 (50.39)
	MSGA	38.90 (2.15)	0.8997 (0.0181)	1.0871 (0.0227)	110.32 (17.32)
	RM	42.21 (0.52)	0.9207 (0.0220)	0.7241 (0.0114)	75.35 (10.87)
	MSMS	47.10 (0.94)	0.9660 (0.0087)	0.3421 (0.0171)	21.74 (8.61)

Maximum number of function evaluations = 2.5×10^4

exploration and exploitation), in a more effective and efficient way than other multimodal search methods by using an acceptable number of function evaluations.

On the other hand, according to the *EPN* values from Table 7.5, we observe that MSMS can always find more optimal solutions for the multimodal problems f_8–f_{14}. For function f_8, only MSMS can find all optima, whereas CP, AEGA, CSA and AiNet exhibit the worst *EPN* performance. A set of special cases are the functions f_9–f_{12} which contain a few prominent optima (with good fitness value); however, such functions present several optima with bad fitness values. In these functions, MSMS is able to detect the highest number of optimum points. On the contrary, most of algorithms are able to find only outstanding solutions. For function f_{13}, four algorithms (CDE, SDE, CP, MGSA, RM and MSMS) can get all optima for each execution. In case of function f_{14}, it features numerous optima with different fitness values. However, MSMS still can find all global optima with an effectiveness rate of 100%. Regarding to the *MPR*, MSMS has reached the best score for all the multimodal problems. On the other hand, the other algorithms show different

Table 7.5 Results of functions f_8–f_{14} of Table 7.10 with $n = 2$

Function	Algorithm	*EPN*	*MPR*	*PA*	*DA*
f_8	CDE	22.25 (1.31)	0.9303 (0.0165)	4.842 (0.624)	0.8945 (0.0643)
	SDE	17.21 (0.85)	0.4298 (0.0184)	21.54 (8.251)	2.327 (0.0571)
	CP	8.00 (2.10)	0.2007 (0.0101)	57.01 (10.21)	6.871 (0.1410)
	AEGA	14.22 (1.77)	0.3794 (0.0382)	36.11 (8.341)	3.411 (0.1821)
	CSA	14.01 (1.07)	0.3002 (0.0102)	40.02 (3.429)	3.6614 (0.3208)
	AiNet	16.51 (1.08)	0.4107 (0.0240)	23.01 (6.851)	2.9961 (0.0541)
	MSGA	20.32 (1.53)	0.8987 (0.0247)	5.32 (0.1241)	1.212 (0.0215)
	RM	23.03 (0.50)	0.9601 (0.0117)	2.872 (0.1740)	0.7454 (0.0171)
	MSMS	25 (0)	0.9960 (0.0044)	1.5041 (0.5645)	0.3601 (0.0120)
f_9	CDE	2.0 (0.18)	0.7781 (0.0132)	23.443 (2.431)	3.0021 (0.1130)
	SDE	2.1 (0.41)	0.8008 (0.0227)	21.104 (3.88)	2.854 (0.6325)
	CP	2.2 (0.11)	0.8321 (0.0417)	18.968 (2.395)	2.792 (0.5201)
	AEGA	2.0 (0.20)	0.7790 (0.0174)	23.228 (3.871)	3.0337 (0.1960)
	CSA	2 (0)	0.7101 (0.0047)	23.337 (7.325)	3.0556 (0.4401)
	AiNet	2 (0)	0.7122 (0.0031)	23.578 (3.251)	3.1143 (0.2213)
	MSGA	3.5 (0.75)	0.8287 (0.0108)	10.241 (2.231)	1.532 (0.0251)
	RM	4.1 (0)	0.9002 (0.0081)	5.214 (0.8721)	0.8874 (0.0047)
	MSMS	5 (0)	0.9900 (0.0102)	1.021 (0.773)	0.012 (0.0096)
f_{10}	CDE	4.02 (0.23)	0.7221 (0.0114)	3.421 (0.124)	3.058 (0.1125)
	SDE	4.15 (0.21)	0.7822 (0.0297)	3.107 (0.871)	2.781 (0.1270)
	CP	4 (1.1)	0.7011 (0.0110)	3.992 (0.987)	3.423 (0.1580)
	AEGA	3.3 (0.5)	0.6689 (0.0121)	4.4101 (0.1485)	4.007 (0.2107)
	CSA	3.52 (0.26)	0.6804 (0.0551)	4.011 (0.1100)	3.632 (0.2020)
	AiNet	4 (0.1)	0.7079 (0.0113)	3.5643 (0.2401)	3.490 (0.1953)
	MSGA	6.10 (0.40)	0.8621 (0.0147)	2.247 (0.134)	1.854 (0.2124)
	RM	6.92 (0.25)	0.9001 (0.0228)	1.005 (0.020)	1.108 (0.0227)
	MSMS	8 (1.1)	0.9800 (0.0501)	0.5871 (0.068)	0.3921 (0.0203)
f_{11}	CDE	10.20 (1.31)	0.8500 (0.0571)	1.897 (0.064)	0.5854 (0.0118)
	SDE	10.12 (1.04)	0.8307 (0.0571)	1.972 (0.087)	0.6721 (0.0337)
	CP	9.0 (0.81)	0.7521 (0.0741)	2.478 (0.074)	0.6927 (0.0287)
	AEGA	8.20 (1.02)	0.6900 (0.0971)	3.897 (0.547)	0.7099 (0.0337)
	CSA	8 (0.2)	0.6520 (0.0741)	4.257 (0.347)	0.7387 (0.0138)
	AiNet	8 (0.1)	0.6477 (0.0851)	4.472 (0.472)	0.7797 (0.0227)
	MSGA	9.3 (0.7)	0.8821 (0.0225)	2.107 (0.055)	0.6854 (0.0115)
	RM	10.20 (0.7)	0.9087 (0.0330)	1.880 (0.067)	0.5721 (0.0224)
	MSMS	12 (0)	0.9914 (0.0041)	0.5742 (0.088)	0.0897 (0.0114)
f_{12}	CDE	6.33 (1.21)	0.6999 (0.0221)	4.011 (0.101)	5.1041 (0.0411)
	SDE	5.21 (1.04)	0.5788 (0.0141)	4.998 (0.174)	5.8841 (0.0112)
	CP	6 (0.33)	0.6521 (0.0874)	4.507 (0.957)	5.3197 (0.1421)
	AEGA	4 (0.5)	0.4100 (0.0147)	6.387 (0.147)	7.8787 (1.0101)

(continued)

Table 7.5 (continued)

Function	Algorithm	*EPN*	*MPR*	*PA*	*DA*
	CSA	4 (0.2)	0.3999 (0.0128)	6.177 (0.876)	7.7988 (1.0100)
	AiNet	4 (0)	0.4033 (0.0543)	6.022 (0.654)	7.6681 (1.0021)
	MSGA	8.4 (1.6)	0.8002 (0.0438)	2.874 (0.067)	3.874 (0.0151)
	RM	10.1 (0.8)	0.9032 (0.0239)	1.587 (0.221)	2.148 (0.0150)
	MSMS	11.60 (0.51)	0.9890 (0.0101)	0.487 (0.097)	0.891 (0.0177)
f_{13}	CDE	13 (0)	1 (0)	0.012 (0.010)	0.040 (0.0112)
	SDE	13 (0)	1 (0)	0.021 (0.008)	0.029 (0.0043)
	CP	13 (0)	1 (0)	0.023 (0.007)	0.052 (0.0088)
	AEGA	10.22 (0.81)	0.8157 (0.0140)	0.097 (0.008)	0.102 (0.0014)
	CSA	8.90 (1.04)	0.7900 (0.0127)	0.128 (0.061)	0.154 (0.0058)
	AiNet	10.16 (0.64)	0.8204 (0.0118)	0.108 (0.024)	0.127 (0.0017)
	MSGA	13 (0)	1 (0)	0.022 (0.006)	0.038 (0.0006)
	RM	13 (0)	1 (0)	0.018 (0.002)	0.033 (0.0009)
	MSMS	13 (0)	1 (0)	0.010 (0.003)	0.011 (0.0005)
f_{14}	CDE	3 (0.2)	0.6607 (0.0150)	0.851 (0.075)	162.24 (10.54)
	SDE	3.25 (0.4)	0.7000 (0.0224)	0.689 (0.011)	122.31 (11.31)
	CP	2.6 (0.5)	0.6102 (0.0129)	1.127 (0.087)	187.01 (21.87)
	AEGA	3 (0.1)	0.6647 (0.0150)	0.920 (0.028)	160.11 (21.18)
	CSA	3 (0.4)	0.6604 (0.0117)	0.987 (0.012)	165.12 (23.11)
	AiNet	3.3 (0.12)	0.7022 (0.0255)	0.627 (0.024)	118.31 (18.11)
	MSGA	5.1 (0.3)	0.8002 (0.0033)	0.257 (0.054)	22.571 (4.21)
	RM	6 (0.4)	0.9021 (0.0210)	0.1007 (0.043)	15.331 (2.81)
	MSMS	8 (0)	0.9974 (0.0061)	0.0814 (0.0010)	9.22 (1.70)

Maximum number of function evaluations = 2.5×10^4

accuracy levels. A close inspection of Table 7.5 also reveals that the MSMS approach is able to achieve the smallest *PA* and *DA* indexes in relation to all other techniques. To statistically examine the results of Tables 7.4 and 7.5, a non-parametric test known as the Wilcoxon analysis [74, 75] has been conducted. It permits to evaluate the differences between two related methods. The test is performed for the 5% (0.005) significance level over the "effective peak number (*EPN*)" data. Table 7.6 reports the *p*-values generated by Wilcoxon analysis for the pair-wise comparison among the algorithms. Under such conditions, eight groups are produced: MSMS versus CDE, MSMS versus SDE, MSMS versus CP, MSMS versus AEGA, MSMS versus CSA, MSMS versus AiNet, MSMS versus MGSA and MSMS versus RM. In the Wilcoxon analysis, it is considered as a null hypothesis that there is no a notable difference between the two methods. On the other hand, it is admitted as alternative hypothesis that there is an important difference between both approaches. In order to facilitate the analysis of Table 7.6, the symbols ▲, ▼, and ▶ are adopted. ▲ indicates that the MSMS method performs significantly better than the tested algorithm on the specified function. ▼ symbolizes that the MSMS algorithm performs worse than the tested algorithm, and ▶ means that the Wilcoxon rank sum test cannot distinguish between the simulation results of the MSMS multimodal optimizer and the tested algorithm. The number of cases that fall in these situations are shown at the bottom of the table.

According to the results of Table 7.6, most of the p-values are less than 0.05 (5% significance level) which is contrary to the null hypothesis and indicates that the MSMS performs better than the other methods. Such data are statistically significant and show that they have not occurred by coincidence (i.e. due to the normal noise existent in the process). From Table 7.6, it is clear that the p-values of functions f_1, f_2, f_6 and f_{13} in the groups MSMS versus CDE, MSMS versus SDE, MSMS versus CP, MSMS versus AEGA, MSMS versus MGSA and MSMS versus RM are higher than 0.05. Such results reveal that there is not statistically difference in terms of precision between both methods that have been applied to the aforementioned functions. Contrarily, according to results of Table 7.6, the MSMS exhibits better *EPN* indexes with regard to CSA and AiNet over all functions.

7.4.3 Comparing MSMS Performance for Composition Functions (High Dimensional)

Benchmark test functions are needed to objectively evaluate the efficacy of a new approach. Different to global search strategies, in multimodal optimization, test functions must provide a set of global and local optima with their precise localization. Therefore, not any benchmark function can be used for these purposes [21, 76]. Most of the test functions utilized in the literature are relatively simple due to their low dimensions. To extend the analysis of multimodal techniques, the composition functions have been recently proposed [21, 76]. Composition functions are multidimensional problems which are constructed as a weighted aggregation of basic functions. As a consequence of this combination, the number of optimal solutions and their positions are easily attainable.

This section presents the performance comparison for different algorithms solving the composite functions F_1–F_4 that are shown in Table 7.11 for $n = 30$. The aim is to determine whether MSMS is more efficient and effective than other existing algorithms for finding all multiple optima. All the algorithms employ a population 50 individuals by using 5×10^4 function evaluations. This stop criterion has been decided to keep compatibility with similar works published in the literature.

Table 7.7 presents the summarized performance results among the algorithms, for functions F_1–F_4. The tables report the comparison in terms of the effective peak number (*EPN*), the maximum peak ratio (*MPR*), the peak accuracy (*PA*) and the distance accuracy (*DA*). The results are analyzed by considering 50 different executions. The goal of multimodal optimizers is to detect as many as possible optima.

The main objective in this experiment is to determine whether MSMS is more efficient and effective than other existing algorithms when it faces multidimensional test functions. In order to statistically analyze the results of Table 7.7, the Wilcoxon analysis has been conducted. Its results are presented in Table 7.8 for the "effective peak number (*EPN*)" index.

After an analysis of Table 7.7, it is clear that the performance of all methods in high-dimensional functions is worse than those presented in low-dimensional problems. This fact represents the difficulty of each method for detecting and maintaining multiple optima under several dimensions. According to Table 7.7, the MSMS approach provides better performance than CDE, SDE, CP, AEGA, CSA, AiNet, MGSA and RM in most of the composition functions in terms of the indexes *EPN*, *MPR*, *PA* and *DA*. The exception is function F_2 where the RM algorithm present the best performance. From the *EPN* measure, we observe that MSMS could always find more optimal solutions for the composition problems F_1–F_4. For most of the functions, MSMS can find most of the optima, whereas CDE, SDE, CP, AEGA, CSA, AiNet and MGSA exhibit the worst *EPN* performance. On the other hand, in general, RM maintain an acceptable performance in comparison to the MSMS method. In terms of number of the *MPR* index, MSMS has obtained the best score for most of the composition problems. The rest of the algorithms present lower accuracy levels. A close inspection of Table 7.7 also reveals that the MSMS approach is able to achieve the smallest *PA* and *DA* indexes in relation to all other methods.

According to the results of Table 7.8, most of the p-values are less than 0.05 (5% significance level) which is against the null hypothesis and indicates that the MSMS performs better than the other methods. In case of function F_2, a close inspection of Table 7.8 shows that the algorithm RM statistically performs better than the MSMS method. Similarly, in function F_3, the Wilcoxon test reveals that there is not statistically difference in terms of precision between RM and MSMS.

7.4.4 Diversity and Exploration

The process of finding so many optima as possible is known as multimodal optimization. In order to attain this objective, one important characteristic of a multimodal optimization algorithm is to conserve the population diversity. This diversity can be interpreted as the efficiency of the exploration search strategy. In this section, the exploration capacities of the multimodal methods is analyzed. For the study, an experiment that evaluates how the population diversity evolves along the search is conducted.

To evaluate the exploration capacities, the diversity of each proposal is analyzed. In the experiment, the diversity defined in Eq. 7.21 is calculated and reported during the optimization of a determined benchmark function. In the comparison, the composition function F_4 is considered. This function has been selected for being representative of the different behaviors presented in multimodal optimization.

Table 7.6 p-values produced by Wilcoxon's test comparing MSMS versus CDE, MSMS versus SDE, MSMS versus CP, MSMS versus AEGA, MSMS versus CSA, MSMS versus AiNet, MSMS versus MGSA and MSMS versus RM over the "effective peak number (EPN)" values from Tables 7.4 and 7.5

MSMS versus	CDE	SDE	CP	AEGA	CSA	AiNet	MGSA	RM
f_1	0.165►	0.154►	0.148►	0.159►	0.033▲	0.039▲	0.168►	0.175►
f_2	0.185►	0.178►	0.167►	0.147►	0.051►	0.056►	0.167►	0.191►
f_3	0.048▲	0.011▲	0.010▲	0.019▲	0.015▲	0.016▲	0.031▲	0.041▲
f_4	0.006▲	0.003▲	0.002▲	0.004▲	0.002▲	0.001▲	0.039▲	0.051►
f_5	0.014▲	0.010▲	0.011▲	0.008▲	0.003▲	0.008▲	0.024▲	0.031▲
f_6	0.180►	0.163►	0.191►	0.090►	0.038▲	0.040▲	0.195►	0.197►
f_7	0.006▲	0.003▲	0.002▲	0.003▲	0.004▲	0.004▲	0.011▲	0.019▲
f_8	0.033▲	0.029▲	0.002▲	0.006▲	0.005▲	0.008▲	0.031▲	0.040▲
f_9	0.008▲	0.012▲	0.011▲	0.012▲	0.006▲	0.004▲	0.036▲	0.042▲
f_{10}	0.017▲	0.019▲	0.021▲	0.020▲	0.011▲	0.016▲	0.026▲	0.031▲
f_{11}	0.017▲	0.019▲	0.011▲	0.014▲	0.008▲	0.004▲	0.021▲	0.028▲
f_{12}	0.013▲	0.015▲	0.018▲	0.004▲	0.011▲	0.009▲	0.012▲	0.039▲
f_{13}	0.207►	0.214►	0.234►	0.087►	0.024▲	0.016▲	0.224►	0.254►
f_{14}	0.003▲	0.006▲	0.009▲	0.008▲	0.007▲	0.003▲	0.009▲	0.011▲
▲	10	10	10	10	13	13	10	9
►	4	4	4	4	1	1	4	5
▼	0	0	0	0	0	0	0	0

Figures 7.6 and 7.7 show the evolution of the diversity under 500 different iterations. In axis x there is the number of evaluations, and in axis y the diversity measure.

After an analysis of the Figs. 7.6 and 7.7, it is clear that all algorithms begin with a big diversity as a consequence of their random initialization. As the iterations increase, the population diversity diminishes. Under such conditions, the MSMS reaches the best population diversity, since it maintains the higher diversity values. The RM method is also a method that obtains a good performance in terms of population diversity, but with values lightly lower than MSMS. The rest of the algorithms maintain almost the same diversity indexes. The good diversity characteristics of the MSMS algorithm observed in this experiments permit affirm that the incorporation of the memory endows of interesting exploration capacities. This fact is a consequence of the high diversity of the solutions contained in the memory **M**.

Table 7.7 Results of functions f_8–f_{14} of Table 7.10 with $n = 30$

Function	Algorithm	*EPN*	*MPR*	*PA*	*DA*
F_1	CDE	3.1 (0.4)	0.5874 (0.0149)	573.295 (89.23)	2.9846 (0.0501)
	SDE	2.8 (0.2)	0.5306 (0.0356)	652.217 (56.34)	3.3955 (0.0286)
	CP	2.7 (0.5)	0.5116 (0.0274)	678.617 (68.10)	3.5329 (0.0394)
	AEGA	3.3 (0.3)	0.6289 (0.0191)	515.632 (76.49)	2.6844 (0.0485)
	CSA	2.5 (0.4)	0.4737 (0.0304)	731.278 (53.48)	3.8071 (0.0370)
	AiNet	2.2 (0.1)	0.4169 (0.0230)	810.201 (101.21)	4.2180 (0.0514)
	MSGA	3.3 (0.3)	0.6253 (0.0117)	520.634 (93.31)	2.7104 (0.0142)
	RM	3.9 (0.5)	0.7390 (0.0228)	362.651 (47.91)	1.8880 (0.0231)
	MSMS	5.2 (0.7)	0.9754 (0.0123)	120.286 (10.36)	0.1056 (0.0158)
F_2	CDE	3.5 (0.2)	0.4547 (0.0154)	628.786 (94.21)	4.0432 (0.1421)
	SDE	3.8 (0.3)	0.4937 (0.0221)	700.142 (53.11)	3.7540 (0.1822)
	CP	3.2 (0.7)	0.4157 (0.0347)	808.022 (66.27)	4.3323 (0.2150)
	AEGA	4.6 (0.8)	0.5976 (0.0187)	556.462 (88.01)	2.9836 (0.3172)
	CSA	2.8 (0.6)	0.3638 (0.0235)	875.524 (93.10)	4.7172 (0.2709)
	AiNet	2.6 (0.9)	0.3378 (0.0362)	915.730 (80.93)	4.9100 (0.3895)
	MSGA	5.7 (0.7)	0.7406 (0.0397)	358.713 (47.30)	1.9233 (0.1206)
	RM	7.6 (0.5)	0.9875 (0.0400)	17.700 (4.21)	0.0926 (0.0101)
	MSMS	6.8 (0.8)	0.8835 (0.0226)	89.534 (10.24)	0.2638 (0.0961)
F_3	CDE	2.1 (0.8)	0.3613 (0.0299)	943.912 (131.40)	5.2739 (0.2711)
	SDE	2.5 (0.6)	0.4302 (0.0334)	842.087 (56.67)	4.7049 (0.1801)
	CP	2.8 (0.9)	0.4818 (0.0412)	364.92 (69.28)	4.2789 (0.3614)
	AEGA	2.6 (0.3)	0.4474 (0.0228)	816.668 (105.74)	4.5629 (0.5101)
	CSA	1.8 (0.6)	0.3097 (0.0179)	1020.171 (127.01)	5.7001 (0.4082)
	AiNet	2.2 (0.8)	0.3785 (0.0371)	918.492 (87.81)	5.1318 (0.2100)
	MSGA	3.6 (0.4)	0.6195 (0.0250)	562.327 (97.33)	3.1418 (0.1001)
	RM	5.5 (0.7)	0.9591 (0.0127)	91.6521 (18.24)	0.6007 (0.1187)
	MSMS	5.8 (0.3)	0.9781 (0.0214)	82.8079 (10.31)	0.5808 (0.1493)
F_4	CDE	2.7 (1.2)	0.3436 (0.0117)	638.615 (87.32)	5.7126 (0.1102)
	SDE	2.8 (0.9)	0.3563 (0.0221)	626.259 (68.14)	5.6021 (0.1005)
	CP	2.2 (1.5)	0.2799 (0.0340)	700.589 (57.10)	6.2670 (0.1921)
	AEGA	3.2 (0.2)	0.4072 (0.0374)	576.738 (39.98)	5.1591 (0.2128)
	CSA	2.4 (0.7)	0.3054 (0.0411)	675.780 (87.75)	6.0450 (0.3721)
	AiNet	1.8 (0.8)	0.2290 (0.0274)	750.110 (162.10)	6.7100 (0.4101)
	MSGA	4.2 (0.6)	0.5344 (0.0313)	452.984 (63.38)	4.0521 (0.1371)
	RM	6.1 (0.3)	0.7762 (0.0414)	217.736 (50.81)	1.9477 (0.0980)
	MSMS	7.7 (0.4)	0.9799 (0.0127)	59.555 (10.96)	0.1749 (0.0281)

Maximum number of function evaluations = 5×10^4

Table 7.8 *p*-values produced by Wilcoxon's test comparing MSMS versus CDE, MSMS versus SDE, MSMS versus CP, MSMS versus AEGA, MSMS versus CSA, MSMS versus AiNet, MSMS versus MGSA and MSMS versus RM over the "effective peak number (EPN)" values from Table 7.7

MSMS versus	CDE	SDE	CP	AEGA	CSA	AiNet	MGSA	RM
F_1	0.0011▲	0.0037▲	0.0028▲	0.0041▲	0.0057▲	0.0021▲	0.0071▲	0.0098▲
F_2	0.0054▲	0.0064▲	0.0081▲	0.0032▲	0.0017▲	0.0019▲	0.0085▲	0.0091▼
F_3	0.0004▲	0.0008▲	0.0005▲	0.0032▲	0.0007▲	0.0015▲	0.0014▲	0.0612▶
F_4	0.0021▲	0.0002▲	0.0008▲	0.0020▲	0.0009▲	0.0016▲	0.0031▲	0.0063▲
▲	4	4	4	4	4	4	4	2
▶	0	0	0	0	0	0	0	1
▼	0	0	0	0	0	0	0	1

7.4.5 Computational Effort

In this section, the time spent by all methods in the solution of a multimodal optimization problem is evaluated. Random methods are, in general, complex pieces of software with several operators and stochastic branches. Therefore, it is difficult to conduct a complexity analysis from a deterministic perspective. Therefore, the computational time (*CT*) is used in order to evaluate the computational effort. The *CT* exhibits the CPU time invested by an algorithm when it is under operation.

In order to assess the computational effort, an experiment is conducted. In the experiment, the Computational Time (*CT*) is evaluated and compared for each algorithm when they operate over the composition functions F_1–F_4 (high dimensional problems). In the comparison, it is considered that each algorithm reaches 5×10^4 function evaluations. All algorithms have been tested in MatLAB by using a computer with a Pentium-4 2.66-GHZ processor, running Windows 10 operating system over 8 Gb of memory. Table 7.9 presents the obtained *CT* values which are averaged considering 50 independent executions.

According to the *CT* values exhibited in Table 7.9, the algorithms can be ranked in the following order AiNet, CSA, CP, SDE, MSMS, MGSA, CDE, AEGA AND RM. Considering to this list, the MSMS method has approximately the median performance with regard to rest of the methods. It is only surpass for AiNet, CSA, CDE and SDE which maintain the worst performance in terms of precision. Under such conditions, the MSMS shows the best compromise between its computational time and its accuracy.

The comparative low computational effort of MSMS can be attributed to the depuration mechanisms. The constant application of the depuration procedure maintains the memory with the lower number of elements as possible. Under such conditions, the number of computations such distances and other expensive computational operations are considerably reduced in comparison to other approaches.

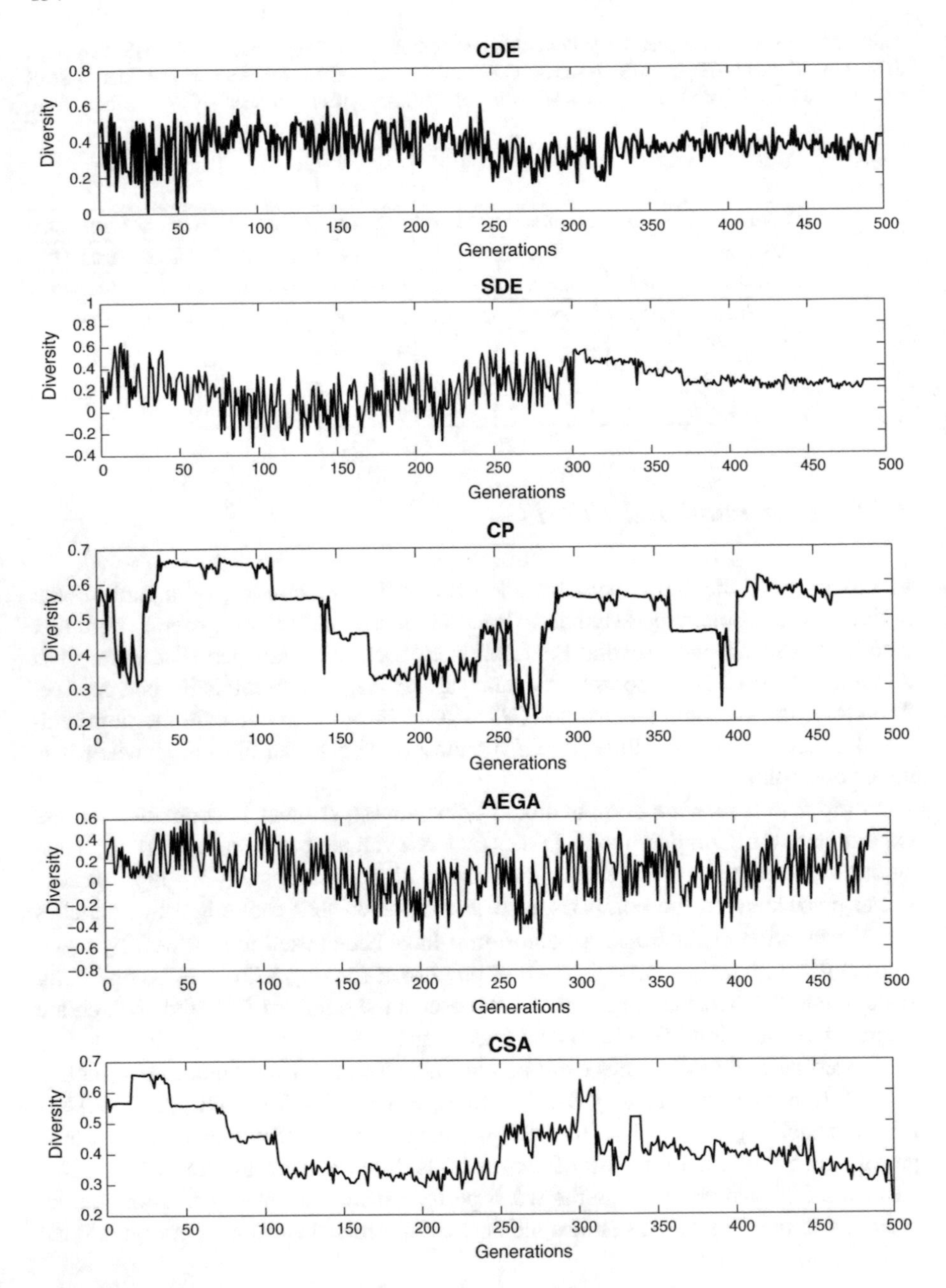

Fig. 7.6 Population diversity evolution for CDE, SDE, CP, AEGA and CSA

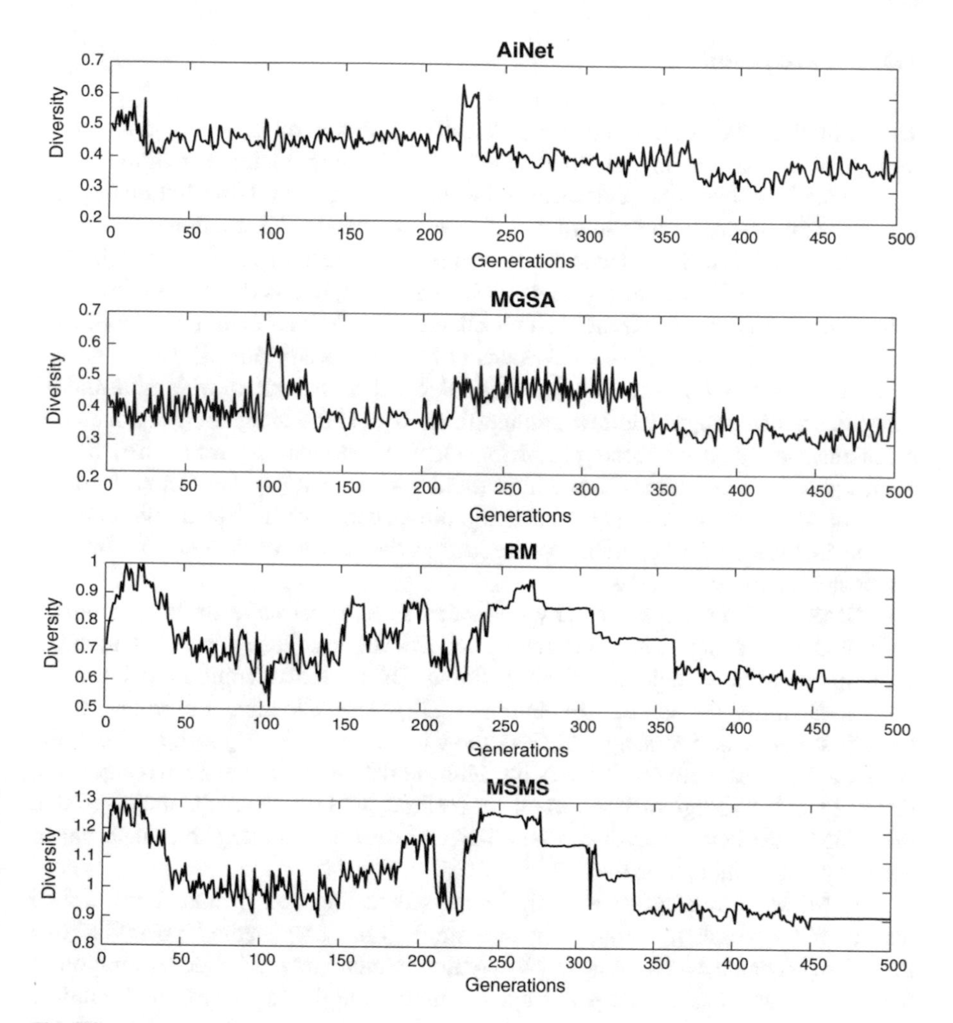

Fig. 7.7 Population diversity evolution for AiNet, MGSA, RM and MSMS

Table 7.9 Computational time (*CT*) values of each algorithm for functions F_1–F_4

	F_1	F_2	F_3	F_4	μ
CDE	23.658s	24.364s	22.843s	25.021s	**23.971s**
SDE	19.104s	22.142s	19.350s	23.154s	**20.937s**
CP	18.534s	20.054s	17.945s	21.271s	**19.449s**
AEGA	24.857s	25.967s	23.952s	26.310s	**24.521s**
CSA	12.082s	14.524s	11.296s	15.532s	**13.358s**
AiNet	11.935s	13.857s	10.533s	14.025s	**12.587s**
MGSA	22.457s	24.058s	21.735s	25.851s	**23.525s**
RM	25.634s	27.004s	24.783s	28.364s	**26.446s**
MSMS	21.085s	22.867s	19.210s	23.932s	**21.773s**

Bold elements respresent the best values

7.5 Conclusions

The main objective of multi-modal optimization is to find multiple global and local optima of a problem in only one single run. Finding multiple solutions to a multi-modal problem is particularly useful in engineering, as the best solution may not always be the best applicable solution due to various practical restrictions. The States of Matter Search (SMS) is a metaheuristic optimization technique. Even though SMS is highly effective in detecting a single global optimum, it fails in providing multiple solutions within a single execution. This chapter introduces a multimodal optimization method called the Multi-modal States of Matter Search (MSMS).

Under the MSMS method, the original SMS is extended with multimodal capacities considering the following modifications: (1) the integration of a memory mechanism to effectively store promising local optima with regard to their fitness values and the separation to other high quality solutions; (2) the alteration of the standard SMS search strategy to accelerate the detection of new local minima; and (3) the inclusion of a depuration procedure at the end of each state to eliminate duplicated memory elements.

MSMS has been experimentally evaluated over a test suite of 18 benchmark multimodal functions. The performance of MSMS has been compared to other existing algorithms including the Crowding Differential Evolution (CDE), the Fitness Sharing Differential Evolution (SDE), the Clearing Procedure (CP), the Elitist-population Strategy (AEGA), the Clonal Selection Algorithm (CSA), the Artificial Immune Network (AiNet), the Multimodal Gravitational Search algorithm (MGSA) and the Region-Based Memetic method (RM). The results indicated that the MSMS method achieves the best balance over its counterparts, in terms of accuracy and computational effort.

The MSMS optimization algorithm allows detecting and maintaining multiple optima for a problem within only one single run. The method possesses two important advantages. (A) It includes operators which allow a better exploration of the search space than other EA approaches, increasing the capacity to find multiple optima. (B) It has the capacity to maintain more and better solutions through the incorporation of an efficient mechanism of memory. However, the MSMS presents an important disadvantage: its implementation is in general more complex than most of the other multimodal algorithms which consider relatively simple operations.

Several research directions could be considered for future work such as the inclusion of other indexes to evaluate similarity between memory elements, the consideration of different probability functions to control the stochastic process, the modification of the evolutionary SMS operators to handle its exploration capacities and the conversion of the optimization procedure into a multi-objective problem.

Appendix: List of Benchmark Functions

See Tables 7.10 and 7.11.

Table 7.10 Low dimensional test functions used in the experimental study

$f(\mathbf{x})$ ($\mathbf{x} = \{x_1, x_2\}$)	S_2	NO	Graph
Bird $f_1(\mathbf{x}) = \sin(x_1) \cdot e^{(1-\cos(x_2))^2} + \cos(x_2) \cdot e^{(1-\sin(x_1))^2} + (x_1 - x_2)^2$	$[-2\pi, 2\pi]$	3	
Cross in tray $f_2(\mathbf{x}) = -0.0001 \cdot \left(\left\| \sin(x_1) \cdot \sin(x_2) \cdot e^{\left\| 100 - \frac{\sqrt{x_1^2 - x_2^2}}{\pi} \right\|} \right\| + 1 \right)^{0.1}$	$[-10, 10]$	12	
DeJongs5 $f_3(\mathbf{x}) = \left\{ 0.002 + \sum_{i=-2}^{2} \sum_{j=-2}^{2} \left[5(i+1) + j + 3 + (x_1 - 16j)^6 + (x_2 - 16i)^6 \right]^{-1} \right\}^{-1}$	$[-40, 40]$	25	
Eggholder $f_4(\mathbf{x}) = -(x_2 + 47) \sin\left(\sqrt{\left\| x_2 + \frac{x_1}{2} + 47 \right\|} \right) - x_1 \sin\left(\sqrt{\left\| x_1 + \frac{x_2}{2} + 47 \right\|} \right)$	$[-512, 512]$	7	

(continued)

Table 7.10 (continued)

$f(\mathbf{x})$ ($\mathbf{x} = \{x_1, x_2\}$)	S_2	NO	Graph
Vincent $f_5(\mathbf{x}) = -\sum_{i=1}^{n} \sin(10 \cdot \log(x_i))$	[0.25,10]	36	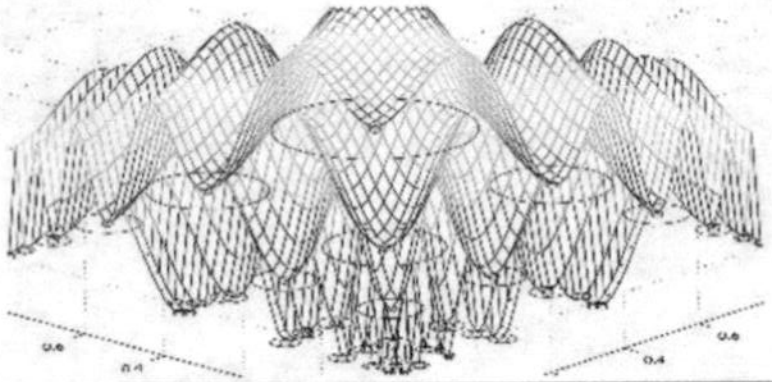
Roots $f_6(\mathbf{x}) = -\left(1 + \left\|(x_1 + x_2 i)^6 - 1\right\|\right)^{-1}$	[−2,2]	6	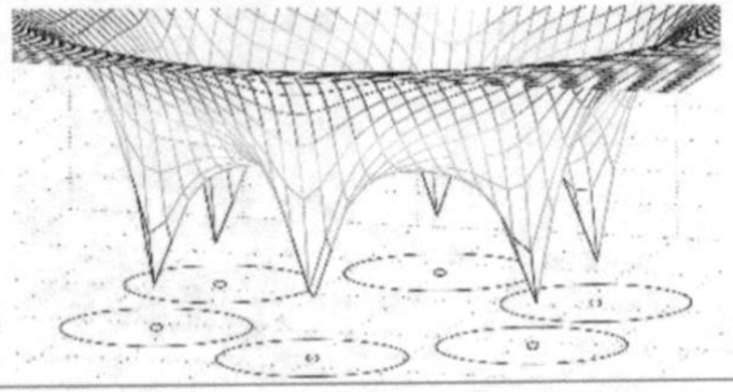
Hilly $f_7(\mathbf{x}) = 10\left[\begin{array}{l} e^{-\frac{\|x_1\|}{50}}\left(1 - \cos\left(\frac{6}{100^{\frac{3}{4}}}\pi\|x_1\|^{\frac{3}{4}}\right)\right) \\ +e^{-\frac{\|x_2\|}{250}}\left(1 - \cos\left(\frac{6}{100^{\frac{3}{4}}}\pi\|x_2\|^{\frac{3}{4}}\right)\right) + 2\left(e^{-\frac{(b-x_1)^2+(b-x_2)^2}{50}}\right) \end{array}\right]$ $with\, b = \left(\frac{5}{6} \cdot 100^{\frac{3}{4}}\right)^{\frac{4}{3}}$	[−100,100]	48	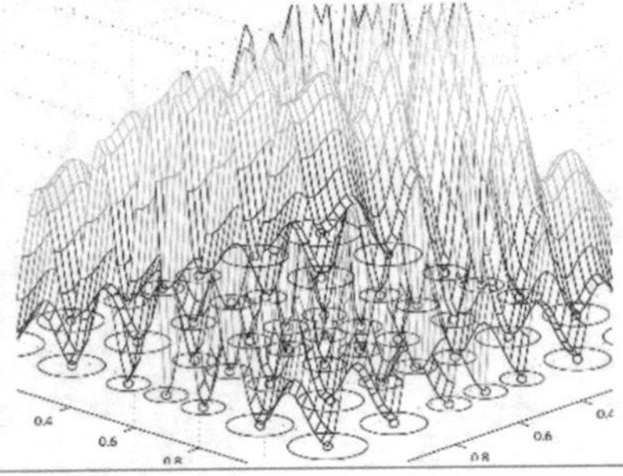

(continued)

Table 7.10 (continued)

$f(\mathbf{x})$ ($\mathbf{x} = \{x_1, x_2\}$)	S_2	NO	Graph
Rastrigin $f_8(\mathbf{x}) = \sum_{i=1}^{n} x_i^2 - 10\cos(2\pi x_i)$	[−5.12,5.12]	25	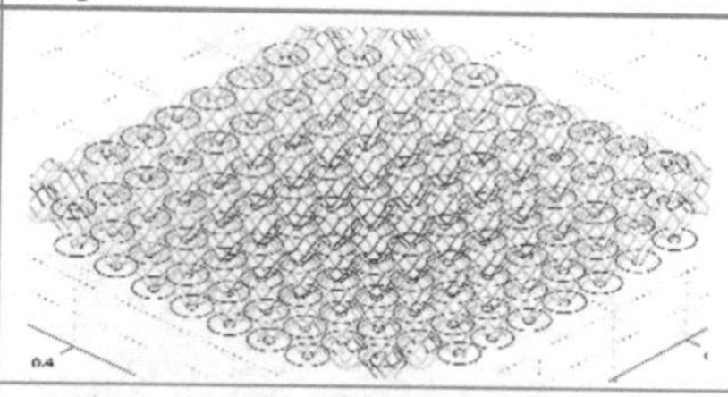
Himmemlblau $f_9(\mathbf{x}) = -\left(x_1^2 + x_2 - 11\right)^2 - \left(x_1 + x_2^2 - 7\right)^2$	[−6,6]	5	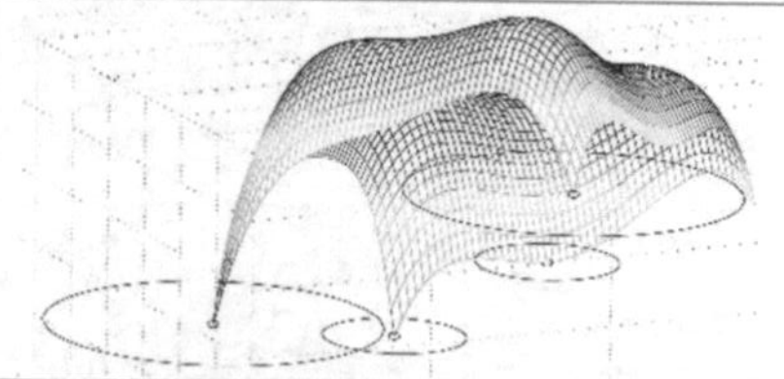
Foxholes $f_{10}(\mathbf{x}) = -\sum_{i=1}^{30}\left(\sum_{j=1}^{n}\left[\left(x_j - a_{ij}\right)^2 + c_j\right]\right)^{-1}$	[0,10]	8	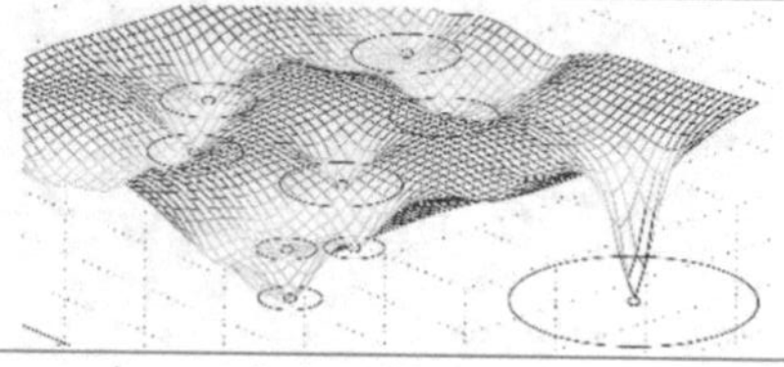
Multimodal F $f_{11}(\mathbf{x}) = -(x_1 \sin(4\pi x_1) - x_2 \sin(4\pi x_2 + \pi))$	[−2,2]	12	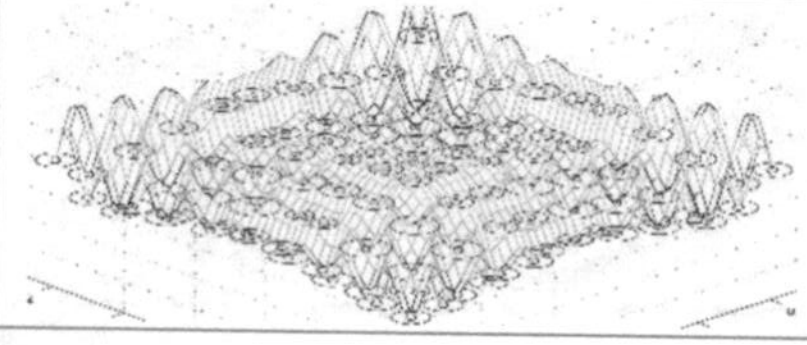

(continued)

Table 7.10 (continued)

$f(\mathbf{x})$ $(\mathbf{x} = \{x_1, x_2\})$	S_2	NO	Graph
Holder Table $f_{12}(\mathbf{x}) = -\sin(x_1)\cos(x_2)e^{\left\lvert 1-\frac{\sqrt{x_1^2+x_2^2}}{\pi}\right\rvert}$	[−10,10]	12	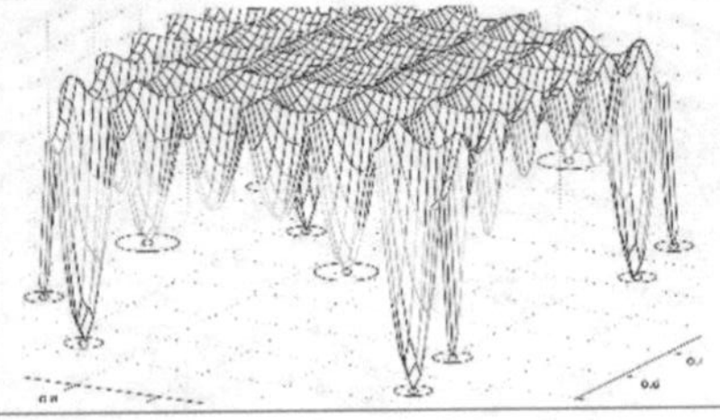
Rastrigin_49m $f_{13}(\mathbf{x}) = \sum_{i=1}^{n} x_i^2 - 18\cos(2\pi x_i)$	[−1,1]	13	
Schwefel $f_{14}(\mathbf{x}) = 418.9829 \cdot n + \sum_{i=1}^{n} -x_i \sin\left(\sqrt{\lvert x_i \rvert}\right)$	[−500,500]	8	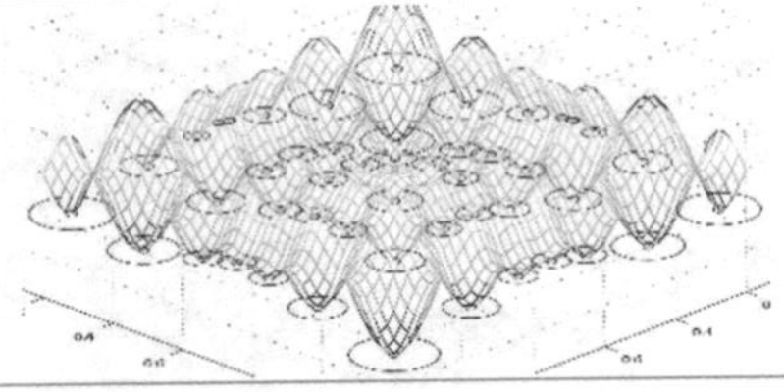

Table 7.11 Composite test functions used in the experimental study

Composition functions	
Basic functions	
Sphere	
$g_1(\mathbf{x}) = \sum_{i=1}^{n} x_i^2$	
Griewank's	
$g_2(\mathbf{x}) = \sum_{i=1}^{n} \frac{x_i^2}{1000} - \prod_{i=1}^{n} \cos\left(\frac{x_i}{\sqrt{i}}\right) + 1$	
Rastringin's	
$g_3(\mathbf{x}) = \sum_{i=1}^{n} (x_i^2 - 10\cos(2\pi x_i) + 10)$	
Weierstrass	
$g_4(\mathbf{x}) = \sum_{i=1}^{n} \left(\sum_{j=0}^{j_{\max}} \kappa^j \cos(2 \cdot \pi \cdot \eta^j (x_i + 0.5))\right) - n \sum_{j=0}^{j_{\max}} \kappa^j \cos(2 \cdot \pi \cdot \eta^j (0.5))$ where $\kappa = 0.5$, $\eta = 3$ and $j_{\max} = 20$	
Expanded Griewank's and Rosenbrock's	
$G(\mathbf{x}) = \sum_{i=1}^{n} \frac{x_i^2}{1000} - \prod_{i=1}^{n} \cos\left(\frac{x_i}{\sqrt{i}}\right) + 1$, $R(\mathbf{x}) = \sum_{i=1}^{n-1} (100(x_i^2 - x_{i+1})^2 + (x_i - 1))$, $g_5(\mathbf{x}) = G(\mathbf{x})R(\mathbf{x})$	
Construction	
A composition function $F_i : S_n \subset \mathbb{R}^n \to \mathbb{R}$ is constructed as a weighted aggregation of h basic functions $u_i : S_n \subset \mathbb{R}^n \to \mathbb{R}$ $(i \in 1, \ldots, h)$. Each basic function $u_i (\in \{g_1, g_2, g_3, g_4, g_5\})$ is shifted to new locations inside the search space S_n and can be either rotated through a linear transformation matrix or used as it is. Therefore, a composition function F_i is built with the following model: $F_i(\mathbf{x}) = \sum_{i=1}^{h} w_i \left(u_i \left(\left(\frac{\mathbf{x} - \mathbf{o}_i}{\lambda_i}\right) \cdot \mathbf{M}_i\right)\right)$, where h represents the number of basic functions used in the composition, g_i denotes the i-th basic function, w_i is the corresponding weight, $\mathbf{o}$ $(\mathbf{o}_i = \{o_i^1, \ldots, o_i^n\})$ is the new shifted optimum of each u_i, $\mathbf{M}_i$ is a linear transformation (rotation) matrix for each u_i, and λ_i is a parameter which is employed to stretch $(\lambda_i > 1)$ or compress $(\lambda_i < 1)$ each function u_i. The weight w_i of each basic function is computed considering the following formulation: $w_i = \exp\left(-\frac{\sum_{j=1}^{n} (x_j - \mathbf{o}_i^j)}{2 \cdot n \cdot \sigma_i^2}\right)$, Then, the weights are recalculated according to: $w_i = \begin{cases} w_i & \text{if } w_1 = \max(w_1) \\ w_i (1 - \max(w_i))^{10} & \text{otherwise} \end{cases}$ The parameter σ_i regulates the coverage range of each basic function g_i, considering a small value of σ_i, it is produced a narrow coverage range for g_i. Finally, the weights are normalized according to $w_i = w_i / \sum_{j=1}^{h} w_j$.	
Definition of composition functions	
Function F_1	Function F_2
$F_1 = \begin{cases} u_1 - u_2 & g_2 \\ u_3 - u_4 & g_4 \\ u_5 - u_6 & g_1 \end{cases}$	$F_2 = \begin{cases} u_1 - u_2 & g_3 \\ u_3 - u_4 & g_4 \\ u_5 - u_6 & g_2 \\ u_7 - u_8 & g_1 \end{cases}$
$h = 6$	$h = 8$

(continued)

Table 7.11 (continued)

$S_n = [-5, 5]^n$	$S_n = [-5, 5]^n$
$\sigma_i = 1, \forall i \in \{1, \ldots, 6\}$	$\sigma_i = 1, \forall i \in \{1, \ldots, 6\}$
$\lambda = [1, 1, 8, 8, 1/5, 1/5]$	$\lambda = [1, 1, 10, 10, 1/10, 1/10, 1/7, 1/7]$
$\mathbf{M}_i = \mathbf{I}, \forall i \in \{1, \ldots, 6\}$	$\mathbf{M}_i = \mathbf{I}, \forall i \in \{1, \ldots, 6\}$
$\mathbf{NO} = 6$	$\mathbf{NO} = 8$
Function F_3	Function F_4
$F_3 = \begin{cases} u_1 - u_2 & g_5 \\ u_3 - u_4 & g_4 \\ u_5 - u_6 & g_2 \end{cases}$	$F_4 = \begin{cases} u_1 - u_2 & g_3 \\ u_3 - u_4 & g_5 \\ u_5 - u_6 & g_4 \\ u_7 - u_8 & g_2 \end{cases}$
$h = 6$	$h = 8$
$S_n = [-5, 5]^n$	$S_n = [-5, 5]^n$
$\sigma = [1, 1, 2, 2, 2, 2]$ $\lambda = [1/4, 1/10, 2, 1, 2, 5]$	$\sigma = [1, 1, 1, 1, 1, 2, 2, 2]$ $\lambda = [4, 1, 4.1, 1/10, 1/5, 1/10, 1/40]$
$\mathbf{M}_i = \mathbf{I}, \forall i \in \{1, \ldots, 6\}$	$\mathbf{M}_i = \mathbf{I}, \forall i \in \{1, \ldots, 6\}$
$\mathbf{NO} = 6$	$\mathbf{NO} = 8$

References

1. Panos, P., Edwin, R., Tuy, H.: Recent developments and trends in global optimization. J. Comput. Appl. Math. **124**, 209–228 (2000)
2. Floudas, C., Akrotirianakis, I., Caratzoulas, S., Meyer, C., Kallrath, J.: Global optimization in the 21st century: advances and challenges. Comput. Chem. Eng. **29**(6), 1185–1202 (2005)
3. Ying, J., Ke-Cun, Z., Shao-Jian, Q.: A deterministic global optimization algorithm. Appl. Math. Comput. **185**(1), 382–387 (2007)
4. Georgieva, A., Jordanov, I.: Global optimization based on novel heuristics, low-discrepancy sequences and genetic algorithms. Eur. J. Oper. Res. **196**, 413–422 (2009)
5. Lera, D., Sergeyev, Y.: Lipschitz and Hölder global optimization using space-filling curves. Appl. Numer. Math. **60**(1–2), 115–129 (2010)
6. Fogel, L.J., Owens, A.J., Walsh, M.J.: Artificial Intelligence Through Simulated Evolution. John Wiley, Chichester, UK (1966)
7. Schwefel, H.P.: Evolution strategies: a comprehensive introduction. J. Nat. Comput. **1**(1), 3–52 (2002)
8. Koza, J.R.: Genetic programming: a paradigm for genetically breeding populations of computer programs to solve problems. Rep. No. STAN-CS-90–1314. Stanford University, CA (1990)
9. Holland, J.H.: Adaptation in Natural and Artificial Systems. University of Michigan Press, Ann Arbor, MI (1975)
10. Goldberg, D.E.: Genetic Algorithms in Search, Optimization and Machine Learning. Addison Wesley, Boston, MA (1989)
11. De Castro, L.N., Von Zuben, F.J.: Artificial immune systems: part I—basic theory and applications. Technical report, TR-DCA 01/99. December 1999

12. Storn, R., Price, K.: Differential evolution-a simple and efficient adaptive scheme for global optimisation over continuous spaces. Technical Report TR-95–012. ICSI, Berkeley, Calif (1995)
13. Kirkpatrick, S., Gelatt, C., Vecchi, M.: Optimization by simulated annealing. Science **220** (4598), 671–680 (1983)
14. İlker, B., Birbil, S., Shu-Cherng, F.: An electromagnetism-like mechanism for global optimization. J. Global Optim. **25**, 263–282 (2003)
15. Rashedia, E., Nezamabadi-pour, H., Saryazdi, S.: Filter modeling using gravitational search algorithm. Eng. Appl. Artif. Intell. **24**(1), 117–122 (2011)
16. Kennedy, J., Eberhart, R.: Particle swarm optimization. In: Proceedings of the 1995 IEEE International Conference on Neural Networks, vol. 4, pp. 1942–1948, December 1995
17. Dorigo, M., Maniezzo, V., Colorni, A.: Positive feedback as a search strategy. Technical Report No. 91-016. Politecnico di Milano (1991)
18. Das, S., Maity, S., Qu, B.Y., Suganthan, P.N.: Real-parameter evolutionary multimodal optimization—a survey of the state-of-the-art. Swarm Evol. Comput. **1**(2), 71–88 (2011)
19. Wong, K.-C., Wu, C.-H., Mok, R.K.P., Peng, C., Zhang, Z.: Evolutionary multimodal optimization using the principle of locality. Inf. Sci. **194**, 138–170 (2012)
20. Tan, K.C., Chiam, S.C., Mamun, A.A., Goh, C.K.: Balancing exploration and exploitation with adaptive variation for evolutionary multi-objective optimization. Eur. J. Oper. Res. **197**, 701–713 (2009)
21. Qu, B.Y., Liang, J.J., Wang, Z.Y., Chen, Q., Suganthan, P.N.: Novel benchmark functions for continuous multimodal optimization with comparative results. Swarm Evol. Comput. **26**, 23–34 (2016)
22. Basak, A., Das, S., Chen-Tan, K.: Multimodal optimization using a biobjective differential evolution algorithm enhanced with mean distance-based selection. IEEE Trans. Evol. Comput. **17**(5), 666–685 (2013)
23. De Jong, K.A.: An analysis of the behavior of a class of genetic adaptive systems. Ph.D. dissertation, University of Michigan, Ann Arbor (1975)
24. Goldberg, D.E., Richardson, J.: Genetic algorithms with sharing for multimodal function optimization. In: Proceedings of 2nd International Conference on Genetic Algorithms, pp. 41–49 (1987)
25. Petrovski, A., Wilson, A., McCall, J.: Statistical analysis of genetic algorithms and inference about optimal factors. Technical Report 2, SCMS Technical Report 1998/2. School of Computer and Mathematical Sciences, Faculty of Science and Technology, The Robert Gordon University, Aberdeen, U.K. (1998)
26. Li, L., Tang, K.: History-based topological speciation for multimodal optimization. IEEE Trans. Evol. Comput. **19**(1), 136–150 (2015)
27. Mengshoel, O.J., Galán, S.F., De Dios, A.: Adaptive generalized crowding for genetic algorithms. Inf. Sci. **258**, 140–159 (2014)
28. Miller, B.L., Shaw, M.J.: Genetic algorithms with dynamic niche sharing for multimodal function optimization. In: Proceedings of the 3rd IEEE Conference on Evolutionary Computation, pp. 786–791 (1996)
29. Thomsen, R.: Multimodal optimization using crowding-based differential evolution. In: Congress on Evolutionary Computation, 2004, CEC2004, vol. 2, pp. 1382–1389
30. Chen, C.-H., Liu, T.-K., Chou, J.-H.: A novel crowding genetic algorithm and its applications to manufacturing robots. IEEE Trans. Ind. Inf. **10**(3), 1705–1716 (2014)
31. Yazdani, S., Nezamabadi-pour, H., Kamyab, S.: A gravitational search algorithm for multimodal optimization. Swarm Evol. Comput. **14**, 1–14 (2014)
32. Chang, W.-D.: A modified particle swarm optimization with multiple subpopulations for multimodal function optimization problems. Appl. Soft Comput. **33**, 170–182 (2015)
33. Liang, J.J., Qu, B.Y., Mao, X.B., Niu, B., Wang, D.Y.: Differential evolution based on fitness Euclidean-distance ratio for multimodal optimization. Neurocomputing **137**, 252–260 (2014)

34. Biswas, S., Das, S., Kundu, S., Patra, G.R.: Utilizing time-linkage property in DOPs: an information sharing based artificial bee colony algorithm for tracking multiple optima in uncertain environments. Soft Comput. **18**, 1199–1212 (2014)
35. Sacco, W.F., Henderson, N., Rios-Coelho, A.C.: Topographical clearing differential evolution: a new method to solve multimodal optimization problems. Prog. Nucl. Energy **71**, 269–278 (2014)
36. Lianga, Y., Kwong-Sak, L.: Genetic algorithm with adaptive elitist-population strategies for multimodal function optimization. Appl. Soft Comput. **11**, 2017–2034 (2011)
37. Gao, W., Yen, G.G., Liu, S.: A cluster-based differential evolution with self-adaptive strategy for multimodal optimization. IEEE Trans. Cybern. **44**(8), 1314–1327 (2014)
38. Qu, B.Y., Suganthan, P.N., Das, S.: A distance-based locally informed particle swarm model for multimodal optimization. IEEE Trans. Evol. Comput. **17**(3), 387–402 (2013)
39. Dong, W., Zhou, M.: Gaussian classier-based evolutionary strategy for multimodal optimization. IEEE Trans. Neural Networks Learn. Syst. **25**(6), 1200–1216 (2014)
40. Hui, S., Suganthan, P.N.: Ensemble and arithmetic recombination-based speciation differential evolution for multimodal optimization. IEEE Trans. Cybern. (In press)
41. Chen, G., Low, C.P., Yang, Z.: Preserving and exploiting genetic diversity in evolutionary programming algorithms. IEEE Trans. Evol. Comput. **13**(3), 661–673 (2009)
42. De Castro, L.N., Zuben, F.J.: Learning and optimization using the clonal selection principle. IEEE Trans. Evol. Comput. **6**, 239–251 (2002)
43. De Castro, L.N., Timmis, J.: An artificial immune network for multimodal function optimization. In: Proceedings of the 2002 IEEE International Conference on Evolutionary Computation, IEEE Press, New York, Honolulu, Hawaii, pp. 699–704 (2002)
44. Xu, Q., Lei, W., Si, J.: Predication based immune network for multimodal function optimization. Eng. Appl. Artif. Intell. **23**, 495–504 (2010)
45. Cuevas, E., González, M.: An optimization algorithm for multimodal functions inspired by collective animal behavior. Soft Comput. **17**(3), 489–502 (2013)
46. Merrikh-Bayat, F.: The runner-root algorithm: a metaheuristic for solving unimodal and multimodal optimization problems inspired by runners and roots of plants in nature. Appl. Soft Comput. **33**, 292–303 (2015)
47. Lacroix, B., Molina, D., Herrera, F.: Region-based memetic algorithm with archive for multimodal optimisation. Inf. Sci. **367–368**, 719–746 (2016)
48. Roya, S., Minhazul, S., Das, S., Ghosha, S., Vasilakos, A.V.: A simulated weed colony system with subregional differential evolution for multimodal optimization. Eng. Optim. **45** (4), 459–481 (2013)
49. Yahyaiea, F., Filizadeh, S.: A surrogate-model based multi-modal optimization algorithm. Eng. Optim. **43**(7), 779–799 (2011)
50. Cuevas, E., Echavarría, A., Ramírez-Ortegón, M.A.: An optimization algorithm inspired by the states of matter that improves the balance between exploration and exploitation. Appl. Intell. **40**(2), 256–272 (2014)
51. Cuevas, E., Echavarría, A., Zaldívar, D., Pérez-Cisneros, M.: A novel evolutionary algorithm inspired by the states of matter for template matching. Expert Syst. Appl. **40**(16), 6359–6373 (2013)
52. Mohamed, A.-A.A., El-Gaafary, A.A.M., Mohamed, Y.S., Hemeida, A.M.: Multi-objective states of matter search algorithm for TCSC-based smart controller design. Electr. Power Syst. Res. **140**, 874–885 (2016)
53. Bailey, R.A.: Association Schemes: Designed Experiments, Algebra and Combinatory. Cambridge University Press, Cambridge (2004)
54. Barr, R.S., Golden, B.L., Kelly, J.P., Resende, M.G., Stewart, W.R.: Designing and reporting on computational experiments with heuristic methods. J Heuristics **1**, 9–32 (1995)
55. Bartz-Beielstein, T.: Experimental research in evolutionary computation—the new experimentalism. In: Natural Computing Series, Springer, Berlin (2006)

56. Batista, E., França, E., Borges, M.: Improving the performance of metaheuristics: an approach combining response surface methodology and racing algorithms. Int. J. Eng. Math. **2015**, Article ID 167031, 9 pages (2015). https://doi.org/10.1155/2015/167031
57. Batista, E., França, E.: Improving the fine-tuning of metaheuristics: an approach combining design of experiments and racing algorithms. J. Optim. **2017**, Article ID 8042436, 7 pages (2017). https://doi.org/10.1155/2017/8042436
58. Calvet, L., Juan, A., Serrat, C., Ries, J.: A statistical learning based approach for parameter fine-tuning of metaheuristics. SORT-Stat. Oper. Res. Trans. **40**(1), 201–224 (2016)
59. Eiben, A.E., Smit, S.K.: Parameter tuning for configuring and analyzing evolutionary algorithms. Swarm Evol. Comput. **1**, 19–31 (2011)
60. Eiben, A.E., Smit, S.K.: Evolutionary algorithm parameters and methods to tune them. In: Monfroy, E., Hamadi, Y., Saubion, F. (eds.) Autonomous Search, pp. 15–36. Springer, Berlin (2012)
61. Karafotias, G., Hoogendoorn, M., Eiben, A.E.: Parameter control in evolutionary algorithms: trends and challenges. IEEE Trans. Evol. Comput. **19**(2), 167–187 (2015)
62. Kok, K.Y., Rajendran, P.: Differential-evolution control parameter optimization for unmanned aerial vehicle path planning. PLoS ONE **11**(3), 1–10 (2016)
63. Ugolotti, R., Cagnoni, S.: Analysis of evolutionary algorithms using multi-objective parameter tuning. In: GECCO '14 Proceedings of the 2014 Annual Conference on Genetic and Evolutionary Computation, pp. 1343–1350
64. Kramer, O., Gloger, B., Gobels, A: An experimental analysis of evolution strategies and particle swarm optimisers using design of experiments. In: GECCO07, pp. 674–681 (2007)
65. Kramer, O.: Evolutionary self-adaptation: a survey of operators and strategy parameters. Evol. Intell. **3**(2), 51–65 (2010)
66. Boari, E., Gisele Pappa, L., Marques, J., Marcos Goncalves, A., Meira, W.: Tuning genetic programming parameters with factorial designs. In: IEEE Congress on Evolutionary Computation (CEC), pp. 1–8 (2010)
67. Czarn, A., MacNish, C., Vijayan, K., Turlach, B., Gupta, R.: Statistical exploratory analysis of genetic algorithms. IEEE Trans. Evol. Comput. **8**(4), 405–421 (2004)
68. Petrovski, A., Brownlee, A, McCall, J.: Statistical optimisation and tuning of GA factors. In: IEEE Congress on Evolutionary Computation, vol. 1, pp. 758–764 (2005)
69. Stodola, P., Mazal, J., Podhorec, M.: Parameter tuning for the ant colony optimization algorithm used in ISR systems. Int. J. Appl. Math. Inform. **9**, 123–126 (2015)
70. Jackson, W., Özcan, E., John, R.: Tuning a simulated annealing metaheuristic for cross-domain search. In: IEEE Congress on Evolutionary Computation 2017, pp. 5–9, Donostia-San Sebastian, Spain (2017)
71. Petrowski, A.: A clearing procedure as a niching method for genetic algorithms. In: Proceedings of the 1996 IEEE International Conference on Evolutionary Computation, pp. 798–803, IEEE Press, New York, Nagoya, Japan (1996)
72. Glover, F.: Tabu search part 1. ORSA J. Comput. **1**(3), 190–206 (1989)
73. Glover, F.: Tabu search part 2. ORSA J. Comput. **1**(3), 4–32 (1990)
74. Garcia, S., Molina, D., Lozano, M., Herrera, F.: A study on the use of non-parametric tests for analyzing the evolutionary algorithms' behaviour: a case study on the CEC '2005, Special session on real parameter optimization. J. Heuristics **15**(6), 617–644 (2009)
75. Wilcoxon, F.: Individual comparisons by ranking methods. Biometrics **1**, 80–83 (1945)
76. Li, X., Engelbrecht, A., Epitropakis, M.G.: Benchmark functions for CEC '2013, Special session and competition on niching methods for multimodal function optimization. Evolutionary Computation (CEC) (2013)

Chapter 8
Metaheuristic Algorithms Based on Fuzzy Logic

Several systems are extremely complicated to be handled quantitatively. In spite of this humans undergo them by using simplistic rules that are obtained from their own experiences. The engineering area that imitates the human reasoning in the use of imprecise information to generate decisions is fuzzy logic. Different to classical methods, fuzzy logic involves a distinct option of processing that allows modeling complex systems using human experience. Recently, many metaheuristic algorithms have been introduced with impressive results. Their operators are based on emulations of natural or social processes to adjust candidate solutions. In this chapter, a methodology to implement human-knowledge as a search strategy is analyzed. In the system, a Takagi-Sugeno inference system is employed to generate a specific search strategy produced by a human specialist. Hence, the number of rules and its effect only depend on the expert experience without considering any learning method. Therefore, each rule describes expert information that models the circumstances under which individuals are adjusted to attain the optimal location.

8.1 Introduction

There are processes that humans can do much better than deterministic systems or computers, such as obstacle avoidance while driving or planning a strategy. This may be due to our unique reasoning capabilities and complex cognitive processing. Although processes can be complex, humans undertake them by using simple rules of thumb extracted from their experiences.

Fuzzy logic [1] is a practical alternative for a variety of challenging applications since it provides a convenient method for constructing systems via the use of heuristic information. The heuristic information may come from a system-operator who has directly interacted with the process. In the fuzzy logic design methodology, this operator is asked to write down a set of rules on how to manipulate the process. We then incorporate these into a fuzzy system that emulates the decision-making

E. Cuevas et al., *Advances in Metaheuristics Algorithms: Methods and Applications*, Studies in Computational Intelligence 775,
https://doi.org/10.1007/978-3-319-89309-9_8

process of the operator [2]. For this reason, the partitioning of the system behavior into regions is an important characteristic of a fuzzy system [3]. In each region, the characteristics of the system can be simply modeled using a rule that associates the region under which certain actions are performed [4]. Typically, a fuzzy model consists of a rule base, where the information available is transparent and easily readable. The fuzzy modeling methodology has been largely exploited in several fields such as pattern recognition [5, 6], control [7, 8] and image processing [9, 10].

Recently, several optimization algorithms based on random principles have been proposed with interesting results. Such approaches are inspired by our scientific understanding of biological or social systems, which at some abstraction level can be represented as optimization processes [11]. These methods mimic the social behavior of bird flocking and fish schooling in the Particle Swarm Optimization (PSO) method [12], the cooperative behavior of bee colonies in the Artificial Bee Colony (ABC) technique [13], the improvisation process that occurs when a musician searches for a better state of harmony in the Harmony Search (HS) [14], the attributes of bat behavior in the Bat Algorithm (BAT) method [15], the mating behavior of firefly insects in the Firefly (FF) method [16], the social behaviors of spiders in the Social Spider Optimization (SSO) [17], the characteristics of animal behavior in a group in the Collective Animal Behavior (CAB) [18] and the emulation of the differential and conventional evolution in species in the Differential Evolution (DE) [19] and Genetic Algorithms (GA) [20], respectively.

On the other hand, the combination of fuzzy systems with metaheuristic algorithms has recently attracted the attention in the Computational Intelligence community. As a result of this integration, a new class of systems known as Evolutionary Fuzzy Systems (EFSs) [21, 22] has emerged. These approaches basically consider the automatic generation and tuning of fuzzy systems through a learning process based on a metaheuristic method. The EFSs approaches reported in the literature can be divided into two classes [21, 22]: tuning and learning.

In a tuning approach, a metaheuristic algorithm is applied to modify the parameters of an existent fuzzy system, without changing its rule base. Some examples of tuning in EFSs include the calibration of fuzzy controllers [23, 24], the adaptation of type-2 fuzzy models [25] and the improvement of accuracy in fuzzy models [26, 27]. In learning, the rule base of a fuzzy system is generated by a metaheuristic algorithm, so that the final fuzzy system has the capacity to accurately reproduce the modeled system. There are several examples of learning in EFSs, which consider different types of problems such as the selection of fuzzy rules with membership functions [28, 29], rule generation [30, 31] and determination of the entire fuzzy structure [32–34].

The analyzed method cannot be considered a EFSs approach, since the fuzzy system, used as optimizer, is not automatically generated or tuned by a learning procedure. On the contrary, its design is based on expert observations extracted from the optimization process. Therefore, the number of rules and its configuration are fixed, remaining static during its operation. Moreover, in a typical EFSs scheme, a metheuristic algorithm is used to find an optimal base rule for a fuzzy system with regard to an evaluation function. Different to such approaches, in our method, a fuzzy system is employed to obtain the optimum value of an optimization problem. Hence, the produced Fuzzy system directly acts as any other metaheuristic algorithm conducting the optimization strategy implemented in its rules.

A metaheuristic algorithm is conceived as a high-level problem-independent methodology that consists of a set of guidelines and operations to develop an optimization strategy. In this chapter, we describe how the fuzzy logic design methodology can be used to construct algorithms for optimization tasks. As opposed to "conventional" metaheuristic approaches where the focus is on the design of optimization operators that emulate a natural or social process, in our approach we focus on gaining an intuitive understanding of how to conduct an efficient search strategy to model it directly into a fuzzy system.

Although sometimes unnoticed, it is well understood that human heuristics play an important role in optimization methods. It must be acknowledged that metaheuristic approaches use human heuristics to tune their corresponding parameters or to select the appropriate algorithm for a certain problem [35]. Under such circumstances, it is important to ask the following questions: How much of the success may be assigned to the use of a certain metaheuristic approach? How much should be attributed to its clever heuristic tuning or selection? Also, if we exploit the use of human heuristic information throughout the entire design process, can we obtain higher performance optimization algorithms?

The use of fuzzy logic for the construction of optimization methods presents several advantages. (A) Generation. "Conventional" metaheuristic approaches reproduce complex natural or social phenomena. Such a reproduction involves the numerical modeling of partially-known behaviors and non-characterized operations, which are sometimes even unknown [36]. Therefore, it is notably complicated to correctly model even very simple metaphors. On the other hand, fuzzy logic provides a simple and well-known method for constructing systems via the use of human knowledge [37]. (B) Transparency. The metaphors used by metaheuristic approaches lead to algorithms that are difficult to understand from an optimization perspective. Therefore, the metaphor cannot be directly interpreted as a consistent search strategy [36]. On the other hand, fuzzy logic generates fully interpretable models whose content expresses the search strategy as humans can conduct it [38]. (C) Improvement. Once designed, metaheuristic methods maintain the same procedure to produce candidate solutions. Incorporating changes to improve the quality of candidate solutions is very complicated and severely damages the conception of the original metaphor [36]. As human experts interact with an optimization process, they obtain a better understanding of the correct search strategies that allow finding the optimal solution. As a result, new rules are obtained so that their inclusion in the existing rule base improves the quality of the original search strategy. Under the fuzzy logic methodology, new rules can be easily incorporated to an already existent system. The addition of such rules allows the capacities of the original system to be extended [39].

In this chapter, a methodology to implement human-knowledge-based optimization strategies is analyzed. In the scheme, a Takagi-Sugeno Fuzzy inference system [40] is used to reproduce a specific search strategy generated by a human expert. Therefore, the number of rules and its configuration only depend on the expert experience without considering any learning rule process. Under these conditions, each fuzzy rule represents an expert observation that models the conditions under which candidate solutions are modified in order to reach the optimal location. To exhibit the performance and robustness of the analyzed method, a comparison to other well-known optimization methods is conducted. The comparison considers several standard benchmark functions which are typically found in the literature of metaheuristic optimization. The results suggest a high performance of the analyzed methodology in comparison to existing optimization strategies.

This chapter is organized as follows: In Sect. 8.2, the basic aspects of fuzzy logic and the different reasoning models are introduced. In Sect. 8.3, the analyzed methodology is exposed. Section 8.4 discusses the characteristics of the analyzed methodology. In Sect. 8.5 the experimental results and the comparative study is presented. Finally, in Sect. 8.6, conclusions are drawn.

8.2 Fuzzy Logic and Reasoning Models

This section presents an introduction to the main fuzzy logic concepts. The discussion particularly considers the Takagi-Sugeno Fuzzy inference model [40].

8.2.1 Fuzzy Logic Concepts

A fuzzy set (A) [1] is a generalization of a Crisp or Boolean set, which is defined in a universe of discourse X. A is a linguistic label which defines the fuzzy set through the word A. Such a word defines how a human expert perceives the variable X in relationship to A. The fuzzy set (A) is characterized by a membership function $\mu_A(x)$ which provides a measure of degree of similarity of an element x from X to the fuzzy set A. It takes values in the interval [0,1], that is:

$$\mu_A(x) : X \rightarrow [0,1] \tag{8.1}$$

Therefore, a generic variable x_c can be represented using multiple fuzzy sets $\{A_1^c, A_2^c, \ldots, A_m^c\}$, each one modeled by a membership function $\left\{\mu_{A_1^c}(x_c), \mu_{A_2^c}(x_c), \ldots, \mu_{A_m^c}(x_c)\right\}$.

A fuzzy system is a computing model based on the concepts of fuzzy logic. It includes three conceptual elements: a rule base, which contains a selection of fuzzy rules; a database, which defines the membership functions used by the fuzzy rules; and a reasoning mechanism, which performs the inference procedure. There are two different inference fuzzy systems: Mamdani [41] and Takagi-Sugeno (TS) [40].

The central difference between the two inference models is in the consequent section of the fuzzy systems. In the Mamdani model, all of the structure of the fuzzy system has linguistic variables and fuzzy sets. However, the consequent section of the TS model consists of mathematical functions. Different to the Mamdani structure, the TS model provides computational efficiency and mathematical simplicity in the rules [42]. Therefore, in order to obtain higher modelling accuracy with fewer rules, the TS fuzzy model is a good candidate that obtains better models when the rules are described as functional associations defined in several local behaviors [42, 43]. Since the available knowledge for the design of the fuzzy system conceived in our approach includes functional, local behaviors, the TS inference model has been used in this work for the system modeling.

8.2.2 *The Takagi-Sugeno (TS) Fuzzy Model*

TS fuzzy systems allow us to describe complicated nonlinear systems by decomposing the input space into several local behaviors, each of which is represented by a simple regression model [3].

The main component of a TS fuzzy system is the set of its K fuzzy rules. They code the human knowledge that explains the performance of the actual process. Each rule denoted by R^i relates the input variables to a consequence of its occurrence. A typical TS fuzzy rule is divided in two parts: Antecedent (I) and consequent (II), which are described as follows:

$$R^i : \overbrace{\text{IF}\, x_1 \text{ is } A_p^1 \text{ and } x_2 \text{ is } A_q^2, \ldots, \text{ and } x_n \text{ is } A_r^n}^{\text{I}} \text{ Then } \underbrace{y_i = g_i(\mathbf{x})}_{\text{II}}, \quad i = 1, 2, \ldots, K \tag{8.2}$$

where $\mathbf{x} = [x_1, x_2, \ldots, x_n]^{\mathrm{T}}$ is the n-dimensional input variable and y_i represents the output rule. $g(\mathbf{x})$ is a function which can be modeled by any function as long as it can appropriately describe the behavior of the system within the fuzzy region specified by the antecedent of rule i. In Eq. 8.2, p, q and r symbolizes one fuzzy set which models the behavior of variables x_1, x_2 and x_n, respectively.

8.2.2.1 Antecedent (I)

The antecedent is a logical combination of simple prepositions of the form "x_e is A_d^e". Such a preposition, modeled by the membership function $\mu_{A_d^e}(x_e)$, provides a measure of degree of similarity between x_e and the fuzzy set A_d^e. Since the antecedent is concatenated by using the "*and*" connector, the degree of fulfilment of the antecedent $\beta_i(\mathbf{x})$ is calculated using a t-norm operator such as the minimum:

$$\beta_i(\mathbf{x}) = \min\left(\mu_{A_p^1}(x_1), \mu_{A_q^2}(x_2), \ldots, \mu_{A_r^n}(x_n)\right) \quad (8.3)$$

8.2.2.2 Consequent (II)

$g_i(\mathbf{x})$ is a function which can be modeled by any function as long as it can appropriately describe the behavior of the system within the fuzzy region specified by the antecedent of rule i.

8.2.2.3 Inference in the TS Model

The global output y of a TS fuzzy system is composed as the concatenation of the local behaviors, and can be seen as the weighted mean of the consequents:

$$y = \frac{\sum_{i=1}^{K} \beta_i(\mathbf{x}) \cdot y_i}{\sum_{i=1}^{K} \beta_i(x)} \quad (8.4)$$

where $\beta_i(\mathbf{x})$ is the degree of fulfillment of the ith rule's antecedent and y_i is the output of the consequent model of that rule. Figure 8.1 shows the fuzzy reasoning procedure for a TS fuzzy system with two rules. The example considers two variables (x_1, x_2) and only two membership functions (I and II) for each variable. Now, it should be clear that the spirit of fuzzy logic systems resembles that of "divide and conquer". Therefore, the antecedent of a fuzzy rule defines a local fuzzy region, while the consequent describes the behavior within the region

Rules

$$R^1 : \text{IF } x_1 \text{ is } A_1^1 \text{ and } x_2 \text{ is } A_2^2 \text{ THEN } y_1 = g_1(\mathbf{x})$$
$$R^2 : \text{IF } x_1 \text{ is } A_2^1 \text{ and } x_2 \text{ is } A_1^2 \text{ THEN } y_2 = g_2(\mathbf{x})$$

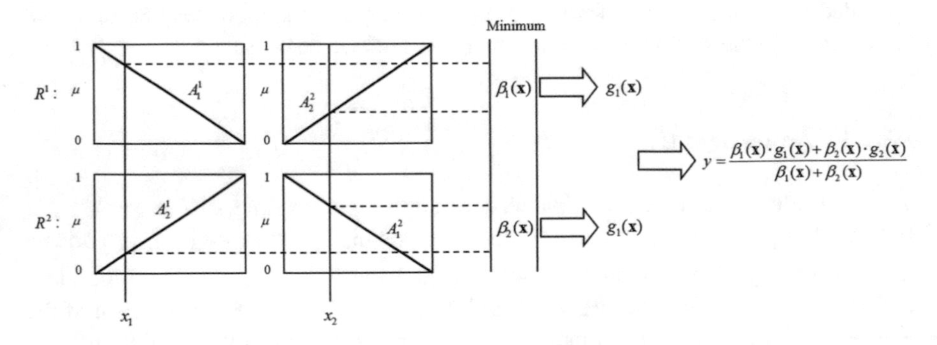

Fig. 8.1 TS fuzzy model

8.3 The Analyzed Methodology

Since there is no specific solution for several kinds of complex problems, human experts often follow a trial-and-error approach to solve them. Under this process, humans obtain experience as the knowledge gained through the interaction with the problem. In general, a fuzzy system is a model that emulates the decisions and behavior of a human that has specialized knowledge and experience in a particular field. Therefore, a fuzzy system is then presumed to be capable of reproducing the behavior of a target system. For example, if the target system is a human operator in charge of a chemical reaction process, then the fuzzy system becomes a fuzzy controller that can regulate the chemical process. Similarly, if the target system is a person who is familiar with optimization strategies and decision-making processes, then the fuzzy inference becomes a fuzzy expert system that can find the optimal solution to a certain optimization problem, as if the search strategy were conducted by the human expert. In this chapter we propose a methodology for emulating human search strategies in an algorithmic structure. In this section, the fuzzy optimization approach is explained in detail. First, each component of the fuzzy system is described; then, the complete computational procedure is presented.

Under a given set of circumstances, an expert provides a description of how to conduct an optimization strategy for finding the optimal solution to a generic problem using natural language. Then, the objective is to take this "linguistic" description and model it into a fuzzy system. The linguistic representation given by the expert is divided into two parts: (A) linguistic variables and (B) rule base formulation.

(A) Linguistic variables describe the way in which a human expert perceives the circumstances of a certain variable in terms of its relative values. One example is the velocity that could be identified as low, moderate and high. (B) Rule base formulation captures the construction process of a set of IF-THEN associations. Each association (rule) expresses the conditions under which certain actions are performed. Typically, a fuzzy model consists of a rule base that maps fuzzy regions to actions. In this context, the contribution of each rule to the behavior of the fuzzy system will be different depending on the operating region.

8.3.1 *Optimization Strategy*

Most of the optimization methods have been designed to solve the problem of finding a global solution to a nonlinear optimization problem with box constraints in the following form [44]:

$$\begin{array}{ll} \text{maximize} & f(\mathbf{x}), \quad \mathbf{x} = (x_1, \ldots, x_n) \in \mathbb{R}^n \\ \text{subject to} & \mathbf{x} \in \mathbf{X} \end{array} \tag{8.5}$$

where $f: \mathbb{R}^n \rightarrow \mathbb{R}$ is a nonlinear function whereas $\mathbf{X} = \{\mathbf{x} \in \mathbb{R}^n | l_i \leq x_i \leq u_i, i = 1,\ldots,n\}$ is a bounded feasible search space, constrained by the lower (l_i) and upper (u_i) limits.

To solve the optimization problem presented in Eq. 8.5, from a population-based perspective [45], a set $\mathbf{P}^k(\{\mathbf{p}_1^k, \mathbf{p}_2^k, \ldots, \mathbf{p}_N^k\})$ of N candidate solutions (individuals) evolves from an initial state ($k = 0$) to a maximum number of generations ($k = Maxgen$). In the first step, the algorithm initiates producing the set of N candidate solutions with values that are uniformly distributed between the pre-specified lower (l_i) and upper (u_i) limits. In each generation, a group of evolutionary operations are applied over the population $\mathbf{P}^k$ to generate the new population $\mathbf{P}^{k+1}$. In the population, an individual $\mathbf{p}_i^k (i \in [1,\ldots,N])$ corresponds to a n-dimensional vector $\left\{p_{i,1}^k, p_{i,2}^k, \ldots, p_{i,n}^k\right\}$ where the dimensions represent the decision variables of the optimization problem to be solved. The quality of a candidate solution $\mathbf{p}_i^k$ is measured through an objective function $f(\mathbf{p}_i^k)$ whose value corresponds to the fitness value of $\mathbf{p}_i^k$. As the optimization process evolves, the best individual $\mathbf{g}$ $(g_1, g_2, \ldots g_n)$ seen so-far is conserved, since it represents the current best available solution.

In the analyzed approach, an optimization human-strategy is modelled in the rule base of a TS Fuzzy inference system, so that the implemented fuzzy rules express the conditions under which candidate solutions from $\mathbf{P}^k$ are evolved to new positions $\mathbf{P}^{k+1}$.

8.3.1.1 Linguistic Variables Characterization (A)

To design a fuzzy system from expert knowledge, it is necessary the characterization of the linguistic variables and the definition of a rule base. A linguistic variable is modeled through the use of membership functions. They represent functions which assign a numerical value to a subjective perception of the variable. The number and the shape of the membership functions that model a certain linguistic variable depend on the application context [46]. Therefore, in order to maintain the design of the fuzzy system as simple as possible, we characterize each linguistic variable by using only two membership functions [47]. One example is the variable velocity V that could be defined by the membership functions: low (μ_L) and high (μ_H). Such membership function are mutually exclusive or disjoint. Therefore, if $\mu_L = 0.7$, then $\mu_H = 0.3$. Assuming that the linguistic variable velocity V has a numerical value inside the interval from 0 to 100 revolutions per minute (rpm), μ_L and μ_H are characterized according to the membership functions shown in Fig. 8.2.

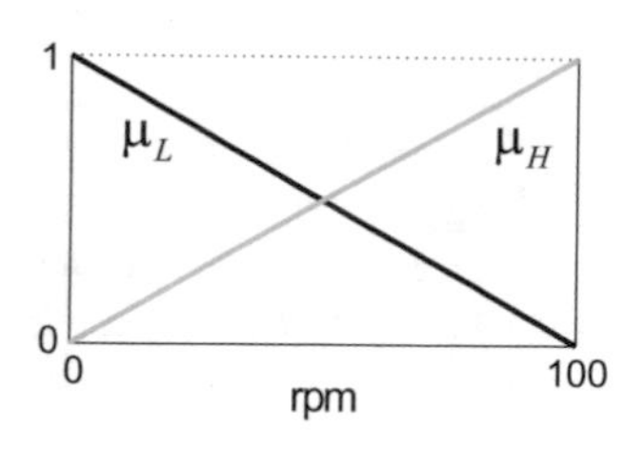

Fig. 8.2 Example of membership functions that characterize a linguistic variable

8.3.1.2 Rule Base Formulation (B)

Several optimization strategies can be formulated by using human knowledge. In this section, a simple search strategy is formulated considering basic observations of the optimization process. Therefore, the simplest search strategy is to move candidate solutions to search regions of the space where it is expected to find the optimal solution. Since the values of the objective function are only known in the positions determined by the candidate solutions, the locations with the highest probabilities of representing potential solutions are those located near the best candidate solution in terms of its fitness value.

Taking this into consideration, a simple search strategy could be formulated by the following four rules:

1. **IF** the distance from $\mathbf{p}_i^k$ to $\mathbf{g}$ is short **AND** $f(\mathbf{p}_i^k)$ is good **THEN** $\mathbf{p}_i^k$ is moved towards (Attraction) $\mathbf{g}$

This rule represents the situation where the candidate solution $\mathbf{p}_i^k$ is moved to the best candidate solution seen so-far $\mathbf{g}$ in order to improve its fitness quality. Since the fitness values of $\mathbf{p}_i^k$ and $\mathbf{g}$ are good in comparison to other members of $\mathbf{P}^k$, the region between $\mathbf{p}_i^k$ and $\mathbf{g}$ maintains promising solutions that could improve $\mathbf{g}$. Therefore, with this movement, it is expected to explore the unknown region between $\mathbf{p}_i^k$ and $\mathbf{g}$. In order to show how each rule performs. Figure 8.3 shows a simple example which expresses the conditions under which action rules are executed. In the example, a population $\mathbf{P}^k$ of five candidate solutions is considered (see Fig. 8.3a). In the case of rule 1, as it is exhibited in Fig. 8.3b, the candidate solution $\mathbf{p}_5^k$ that fulfills the rule requirements is attracted to $\mathbf{g}$

2. **IF** the distance from $\mathbf{p}_i^k$ to $\mathbf{g}$ is short **AND** $f(\mathbf{p}_i^k)$ is bad **THEN** $\mathbf{p}_i^k$ is moved away from (Repulsion) $\mathbf{g}$

In this rule, although the distance between $\mathbf{p}_i^k$ and $\mathbf{g}$ is short, the evidence shows that there are no good solutions between them. Therefore, the improvement of $\mathbf{p}_i^k$ is searched in the opposite direction of $\mathbf{g}$. A visual example of this behavior is presented in Fig. 8.3c

3. **IF** the distance from $\mathbf{p}_i^k$ to $\mathbf{g}$ is large **AND** $f(\mathbf{p}_i^k)$ is good **THEN** $\mathbf{p}_i^k$ is refined

Under this rule, a good candidate solution $\mathbf{p}_i^k$ that is far from $\mathbf{g}$ is refined by searching within its neighborhood. The idea is to improve the quality of competitive candidate solutions which have already been found (exploitation). Such a scenario is presented in Fig. 8.3d where the original candidate solution $\mathbf{p}_2^k$ is substituted by a new position $\mathbf{p}_2^{k+1}$ which is randomly produced within the neighborhood of $\mathbf{p}_2^k$

(continued)

(continued)

4. **IF** the distance from $\mathbf{p}_i^k$ to $\mathbf{g}$ is large **AND** $f(\mathbf{p}_i^k)$ is bad **THEN** a new position is randomly chosen
This rule represents the situation in Fig. 8.3e where the candidate solution $\mathbf{p}_4^k$ is so bad and so far from $\mathbf{g}$ that is better to replace it by other solution ($\mathbf{p}_4^{k+1}$) randomly produced within the search space $\mathbf{X}$

Each of the four rules listed above is a "linguistic rule" which contains only linguistic information. Since linguistic expressions are not well-defined descriptions of the values that they represent, linguistic rules are not accurate. They represent only conceptual ideas about how to achieve a good optimization strategy according to the human perspective. Under such conditions, it is necessary to define the meaning of their linguistic descriptions from a computational point of view.

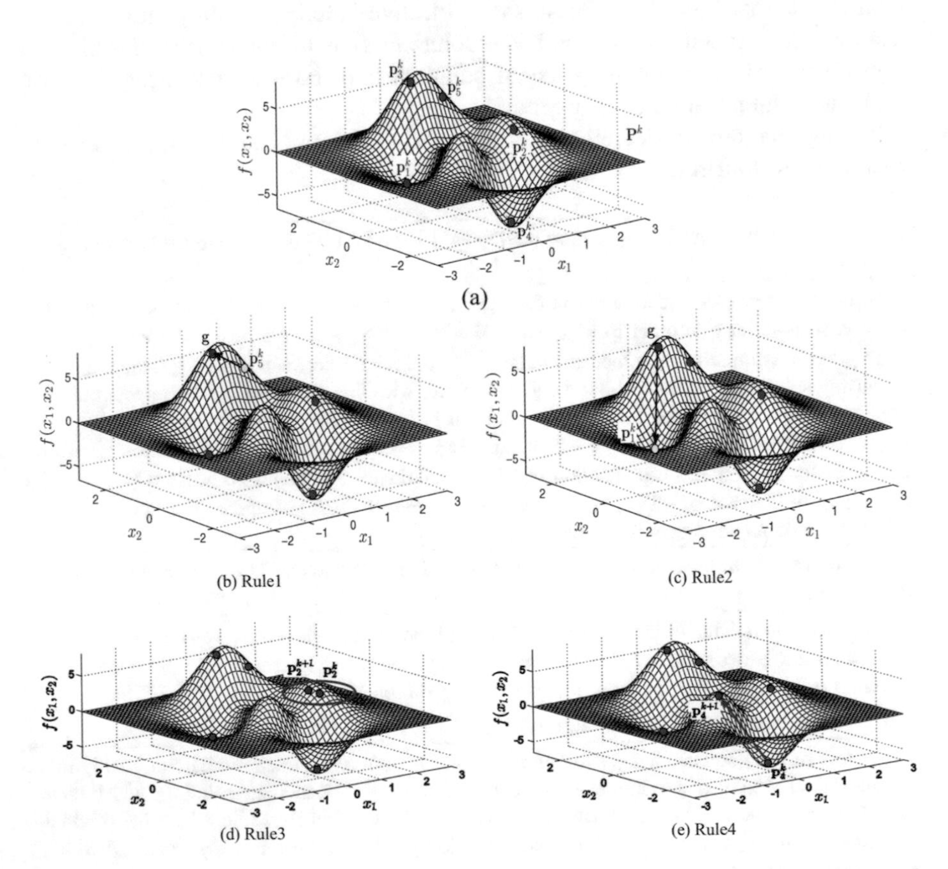

Fig. 8.3 Visual example that expresses the conditions under which action rules are executed. **a** Current configuration of the candidate solution population $\mathbf{P}^k$, **b** rule 1, **c** rule 2, **d** rule 3 and **e** rule 4

8.3.1.3 Implementation of the TS Fuzzy System

In this section, we will discuss the implementation of the expert knowledge concerning the optimization process in a TS fuzzy system.

(I) Membership functions and antecedents

In the rules, two different linguistic variables are considered, distance from de candidate solution $\mathbf{p}_i^k$ to the best solution $\mathbf{g}(D(\mathbf{p}_i^k, \mathbf{g}))$ and the fitness value of the candidate solution $(f(\mathbf{p}_i^k))$. Therefore, $D(\mathbf{p}_i^k, \mathbf{g})$ is characterized by two membership functions: short and large (see 8.3.1.1). On the other hand, $f(\mathbf{p}_i^k)$ is modeled by the membership functions good and bad. Figure 8.4 shows the fuzzy membership functions for both linguistic variables.

The distance $D(\mathbf{p}_i^k, \mathbf{g})$ is defined as the Euclidian distance $\|\mathbf{g} - \mathbf{p}_i^k\|$. Therefore, as it is exhibited in Fig. 8.4a, two complementary membership functions define the relative distance $D(\mathbf{p}_i^k, \mathbf{g})$: short (**S**) and large (**L**). Their support values are 0 and $d_{\max}$, where $d_{\max}$ represents the maximum possible distance delimited by the search space **X** which is defined as follows:

$$d_{\max} = \sqrt{\sum_{s=1}^{d} (u_s - l_s)^2}, \tag{8.6}$$

where d represents the number of dimensions in the search space **X**. In the case of $f(\mathbf{p}_i^k)$, two different membership functions define its relative value: bad (**B**) and good (**G**). Their support values are $f_{\min}$ and $f_{\max}$. These values represent the minimum and maximum fitness values seen so-far. Therefore, they can defined as following:

$$f_{\min} = \min_{\substack{i \in \{1,2,\ldots,N\} \\ k \in \{1,2,\ldots,gen\}}} (f(\mathbf{p}_i^k)) \quad \text{and} \quad f_{\max} = \max_{\substack{i \in \{1,2,\ldots,N\} \\ k \in \{1,2,\ldots,gen\}}} (f(\mathbf{p}_i^k)) \tag{8.7}$$

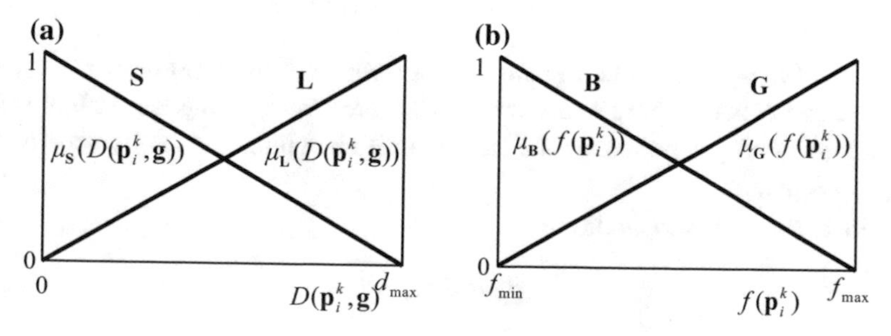

Fig. 8.4 Membership functions for **a** distance $D(\mathbf{p}_i^k, \mathbf{g})$ and **b** for $f(\mathbf{p}_i^k)$

Table 8.1 Degree of fulfilment of the antecedent $\beta_w(\mathbf{x})$ for each rule ($w \in [1, 2, 3, 4]$)

Rule	Degree of fulfilment $\beta_w(\mathbf{x})$
1	$\beta_1(\mathbf{p}_i^k) = \min(\mu_{\mathbf{S}}(D(\mathbf{p}_i^k, \mathbf{g})), \mu_{\mathbf{G}}(f(\mathbf{p}_i^k)))$
2	$\beta_2(\mathbf{p}_i^k) = \min(\mu_{\mathbf{S}}(D(\mathbf{p}_i^k, \mathbf{g})), \mu_{\mathbf{B}}(f(\mathbf{p}_i^k)))$
3	$\beta_3(\mathbf{p}_i^k) = \min(\mu_{\mathbf{L}}(D(\mathbf{p}_i^k, \mathbf{g})), \mu_{\mathbf{G}}(f(\mathbf{p}_i^k)))$
4	$\beta_4(\mathbf{p}_i^k) = \min(\mu_{\mathbf{L}}(D(\mathbf{p}_i^k, \mathbf{g})), \mu_{\mathbf{B}}(f(\mathbf{p}_i^k)))$

From Eq. 8.7, it is evident that $f_{\max} = f(\mathbf{g})$. If a new minimum or maximum value of $f(\mathbf{p}_i^k)$ is detected during the evolution process, it replaces the past values of $f_{\min}$ or $f_{\max}$. Figure 8.4b shows the membership functions that describe $f(\mathbf{p}_i^k)$.

Considering the membership functions defined in Fig. 8.4, the degree of fulfilment of the antecedent $\beta_w(\mathbf{x})$ for each rule ($w \in [1, 2, 3, 4]$) is defined in Table 8.1.

(II) Actions or consequents

Actions or Consequents are functions which can be modeled by any function as long as it can appropriately describe the desired behavior of the system within the fuzzy region specified by the antecedent of a rule i ($i \in [1, 2, 3, 4]$). The consequents of the four rules are modeled by using the following behaviors.

Rule 1. Attraction

$$At(\mathbf{p}_i^k) = \left|f_{\max} - f(\mathbf{p}_i^k)\right| \cdot (\mathbf{g} - \mathbf{p}_i^k) \cdot \alpha_1, \tag{8.8}$$

where α_1 represents a tuning factor. Under this rule, the function $At(\mathbf{p}_i^k)$ produces a change of position in the direction of the attraction vector $(\mathbf{g} - \mathbf{p}_i^k)$. The magnitude depends on the difference of the fitness values between $\mathbf{g}$ and $\mathbf{p}_i^k$.

Rule 2. Repulsion

$$Rep(\mathbf{p}_i^k) = \left|f_{\max} - f(\mathbf{p}_i^k)\right| \cdot (\mathbf{g} + \mathbf{p}_i^k) \cdot \alpha_2, \tag{8.9}$$

where α_2 represents a tuning factor.

Rule 3. Refining or perturbation.

$$Ref(\mathbf{p}_i^k) = \left|f_{\max} - f(\mathbf{p}_i^k)\right| \cdot \mathbf{v} \cdot \gamma, \tag{8.10}$$

where $\mathbf{v} = \{v_1, v_2, \ldots, v_d\}$ is a random vector where each component represents a random number between -1 and 1 whereas γ represents a tuning factor. In this rule, the function $Ref(\mathbf{p}_i^k)$ generates a random position within the limits specified by $\pm\left|f_{\max} - f(\mathbf{p}_i^k)\right|$.

Rule 4. Random substitution.

$$Ran(\mathbf{p}_i^k) = \mathbf{r}, \tag{8.11}$$

where $\mathbf{r} = \{r_1, r_2, \ldots, r_d\}$ is a random vector where each component r_u represents a random number between the lower (l_u) and upper (u_u) limits of the search space $\mathbf{X}$.

(III) Inference of the TS model.

The global change of position $\Delta\mathbf{p}_i^k$ of the TS fuzzy system is composed as the concatenation of the local behaviors produced by the four rules, and can be seen as the weighted mean of the consequents:

$$\Delta\mathbf{p}_i^k = \frac{At(\mathbf{p}_i^k) \cdot \beta_1(\mathbf{p}_i^k) + Rep(\mathbf{p}_i^k) \cdot \beta_2(\mathbf{p}_i^k) + Ref(\mathbf{p}_i^k) \cdot \beta_3(\mathbf{p}_i^k) + Ran(\mathbf{p}_i^k) \cdot \beta_4(\mathbf{p}_i^k)}{\beta_1(\mathbf{p}_i^k) + \beta_2(\mathbf{p}_i^k) + \beta_3(\mathbf{p}_i^k) + \beta_4(\mathbf{p}_i^k)} \tag{8.12}$$

Once $\Delta\mathbf{p}_i^k$ has been calculated, the new position $\mathbf{p}_i^{k+1}$ is calculated as follows:

$$\mathbf{p}_i^{k+1} = \mathbf{p}_i^k + \Delta\mathbf{p}_i^k, \tag{8.13}$$

8.3.2 Computational Procedure

The analyzed algorithm is implemented as an iterative process in which several operations are executed. Such operations can be summarized in the form of pseudo-code in Algorithm 1. The Fuzzy method uses as input information the number of candidate solutions (N), the maximum number of generations ($Maxgen$), and the tuning parameters α_1, α_2, γ. Similar to other metaheuristic algorithms, in the first step (line 2), the algorithm initiates producing the set of N candidate solutions with values that are uniformly distributed between the pre-specified lower and upper limits. These candidate solutions represent the first population $\mathbf{P}^0$. After initialization, the best element $\mathbf{g}$ in terms of its fitness value is selected (line 3). Then, for each particle $\mathbf{p}_i^k$ its distance to the best value $\mathbf{g}$ is calculated (line 6). With $D(\mathbf{p}_i^k, \mathbf{g})$ and $f(\mathbf{p}_i^k)$, the search optimization strategy implemented in the fuzzy system is applied (lines 7–9). Under such circumstances, the antecedents (line 7) and consequents (line 8) are computed while the final displacement $\Delta\mathbf{p}_i^k$ is obtained as a result of the operation performed by the TS model (line 9). Afterwards, the new position $\mathbf{p}_i^{k+1}$ is updated (line 10). Once the new population $\mathbf{P}^{k+1}$ is obtained as a result of the iterative operation of lines 6–10, the best value $\mathbf{g}$ is updated (line 12). This cycle is repeated until the maximum number the iterations $Maxgen$ has been reached.

Algorithm 1. Pseudo-code for the proposed Fuzzy method		
1.	**Input:** *N*, *Maxgen*, α_1, α_2, γ, *k*=0.	
2.	$\mathbf{P}^k \leftarrow$ **Initialize**(*N*);	
3.	**g** $\leftarrow$ **SelectBestParticle(** $\mathbf{P}^k$ **)**;	
4.	**while** *k*<=*Maxgen* **do**	
5.	**for** (*i*=1;*i*>*N*;*i*++)	
6.	$D(\mathbf{p}_i^k, \mathbf{g}) \leftarrow$ **CalculateTheDistancetoTheBest(** $\mathbf{p}_i^k$,**g**);	
7.	$[\beta_1, \beta_2, \beta_3, \beta_4] \leftarrow$ **EvaluateAntecedents(** $D(\mathbf{p}_i^k, \mathbf{g}), f(\mathbf{p}_i^k)$);	Fuzzy System
8.	[*At*,*Rep*,*Ref*,*Ran*] $\leftarrow$ **EvaluateConsequents(** $\mathbf{p}_i^k$,**g**, $f(\mathbf{p}_i^k)$);	
9.	$\Delta\mathbf{p}_i^k \leftarrow$ **InferenceTS(** $\beta_1, \beta_2, \beta_3, \beta_4$,*At*,*Rep*,*Ref*,*Ran*);	
10.	$\mathbf{p}_i^{k+1} \leftarrow \mathbf{p}_i^k + \Delta\mathbf{p}_i^k$	
11.	end **for**	
12.	**g** $\leftarrow$ **SelectBestParticle(** $\mathbf{P}^{k+1}$ **)**;	
13.	$k \leftarrow k+1$	
14.	end **while**	
15.	**Output: g**	

Algorithm 1. Summarized operations of the Fuzzy method

8.4 Discussion About the Methodology

In this section, several important characteristics of the analyzed algorithm are discussed. First, in Sect. 8.4.1, interesting operations of the optimization process are analyzed. Next, in Sect. 8.4.2 the modelling properties of the Fuzzy approach are highlighted.

8.4.1 Optimization Algorithm

A metaheuristic algorithm is conceived as a high-level problem-independent methodology that consists of a set of guidelines and operations to develop an optimization strategy. In the analyzed methodology, a fuzzy system is generated based on expert observations about the optimization process. The final fuzzy system then performs various fuzzy logic operations to produce a new candidate solution $\mathbf{p}_i^{k+1}$ from the current solution $\mathbf{p}_i^k$. During this process, the following operations are involved:

1. Determination of the degree of membership between the input data ($D(\mathbf{p}_i^k, \mathbf{g})$, $f(\mathbf{p}_i^k)$) and the defined fuzzy sets ("short and large" or "good and bad").
2. Calculation of the degree of relevance for each rule based on the degree of fulfilment $\beta_w(\mathbf{x})$ for each rule $(w \in [1, 2, 3, 4])$ in the antecedent part of the rule.
3. Evaluation of the consequent of each rule: *At*, *Rep*, *Ref*, *Ran*.
4. Derivation of the new candidate solution $\mathbf{p}_i^{k+1}$ based on the weighted mean of the consequent functions, according to the TS model.

Under such circumstances, the generated fuzzy system is applied over all candidate solutions from $\mathbf{P}^k$ in order to produce the new population $\mathbf{P}^{k+1}$. This procedure is iteratively executed until a termination criteria has been reached.

8.4.2 *Modeling Characteristics*

Metaheuristic algorithms are widely employed for solving complex optimization problems. Such algorithms have been developed by a combination of deterministic models and randomness, mimicking the behavior of biological or social systems. Most of the metaheuristic methods divide the individual behavior into several processes which show no coupling among them [11, 36].

In the analyzed methodology, the produced fuzzy system models a complex optimization strategy. This modeling is accomplished by a number of fuzzy IF-THEN rules, each of which describes the local behavior of the model. In particular, the rules express the conditions under which new positions are explored. In order to calculate a new candidate solution $\mathbf{p}_i^{k+1}$, the consequent actions of all rules are aggregated. In this way, all the actions are presented in the computation of a certain solution $\mathbf{p}_i^{k+1}$, but with different influence levels. By coupling local behaviors, fuzzy systems are able to reproduce complex global behaviors. An interesting example of such modeling characteristics is rule 1 and rule 2. If these rules are individually analyzed, the attraction and repulsion movements conducted by the functions are completely deterministic. However, when all rules are considered, rule 3 and rule 4 add randomness to the final position of $\mathbf{p}_i^{k+1}$.

8.5 Experimental Results

An illustrative set of 19 functions has been used to examine the performance of our approach. These test functions represent the base functions from the latest competition on single objective optimization problems at CEC2015 [48]. Tables 8.17, 8.18 and 8.19 in Appendix show the benchmark functions employed in our experiments. These functions are ordered into three different classes: Unimodal (Table 8.17), multimodal (Table 8.18) and Hybrid (Table 8.19) test functions.

In the tables, n represents the dimension in which the function is operated, $f(\mathbf{x}^*)$ characterizes the optimal value of the function in the position $\mathbf{x}^*$ and S is the defined search space.

The main objective of this section is to present the performance of the Fuzzy algorithm on numeric optimization problems. Moreover, the results of our method are compared with some popular optimization algorithms by using the complete set of benchmark functions. The results of the Fuzzy algorithm are verified by a statistical analysis of the experimental data.

The experimental results are divided into two sub-sections. In the first section, the performance of the Fuzzy algorithm is evaluated with regard to its tuning parameters. In the second section, the overall performance of the Fuzzy method is compared to six popular optimization algorithms based on random principles.

8.5.1 Performance Evaluation with Regard to Its Own Tuning Parameters

The three parameters of the rules α_1, α_2 and γ affect the expected performance of the fuzzy optimization algorithm. In this sub-section we analyze the behavior of the Fuzzy algorithm considering the different settings of these parameters. All experiments have been executed on a Pentium dual-core computer with 2.53-GHz and 4-GB RAM under MATLAB 8.3. For the sake of simplicity, only the functions from f_1 to f_{14} (unimodal and multimodal) have been considered in the tuning process. In the simulations, all the functions operate with a dimension $n = 30$. As an initial condition, the parameters α_1, α_2 and γ are set to their default values $\alpha_1 = 1.4$, $\alpha_2 = 0.05$ and $\gamma = 0.005$. Then, in our analysis, the three parameters are evaluated one at a time, while the other two parameter remain fixed to their default values. To minimize the stochastic effect of the algorithm, each benchmark function is executed independently a total of 10 times. As a termination criterion, the maximum number of iterations (*Maxgen*) is set to 1000. In all simulations, the population size N is fixed to 50 individuals.

In the first stage, the behavior of the Fuzzy algorithm is analyzed considering different values for α_1. In the analysis, the values of α_1 vary from 0.6 to 1.6 whereas the values of α_2 and γ remain fixed at 0.05 and 0.005, respectively. In the simulation, the Fuzzy method is executed independently 30 times for each value of α_1 on each benchmark function. The results are registered in Table 8.2. These values represent the average best fitness values $(\bar{f})$ and the standard deviations (σ_f) obtained in terms of a certain parameter combination of α_1, α_2 and γ. From Table 8.2, we can conclude that the fuzzy algorithm with $\alpha_1 = 1.4$ maintains the best performance on functions f_1–f_9, and f_{11}. Under this configuration, the algorithm obtains the best results in 9 out of 14 functions. On the other hand, when the parameter α_1 is set to any other value, the performance of the algorithm is inconsistent, producing generally bad results.

Table 8.2 Experimental results obtained by the fuzzy algorithm using different values of α_1

α_1		0.6	0.7	0.8	0.9	1	1.2	1.3	1.4	1.5	1.6
f_1	$\bar{f}$	6.95E−55	7.74E−89	3.97E−167	1.01E−39	1.02E−193	0.00E+00	4.26E−29	**3.08E−281**	1.15E−28	6.85E−28
	σ_f	3.67E−54	4.24E−88	0.00E+00	5.54E−39	0.00E+00	0.00E+00	2.19E−28	**0.00E+00**	3.41E−28	1.08E−27
f_2	$\bar{f}$	6.10E−23	1.14E−53	1.12E−124	2.49E−139	1.08E−158	7.16E−22	1.31E−78	**2.66E−207**	2.03E−15	7.52E+00
	σ_f	3.34E−22	6.07E−53	4.81E−124	1.36E−138	0.00E+00	3.89E−21	7.18E−78	**0.00E+00**	4.31E−15	2.35E+01
f_3	$\bar{f}$	4.69E−10	1.80E−17	2.22E−22	3.23E−22	2.00E−27	4.96E−24	4.11E−27	**1.00E−27**	2.68E−18	1.93E−11
	σ_f	1.62E−09	6.81E−17	8.87E−22	1.64E−21	4.73E−27	2.66E−23	9.21E−27	**1.50E−27**	1.12E−17	5.52E−11
f_4	$\bar{f}$	1.55E−23	1.48E−30	1.51E−130	4.64E−180	2.84E−112	6.13E−19	2.00E−183	**3.85E−220**	9.09E−16	5.16E−15
	σ_f	7.09E−23	8.12E−30	8.25E−130	0.00E+00	1.56E−111	3.31E−18	0.00E+00	**0.00E+00**	2.76E−15	6.91E−15
f_5	$\bar{f}$	2.85E+01	2.85E+01	2.85E+01	2.55E+01	3.85E+01	1.75E+01	2.65E+01	**3.04E−03**	2.85E+01	2.99E+01
	σ_f	4.38E−02	3.86E−02	3.87E−02	4.37E−02	3.04E−02	3.72E−02	4.50E−02	**3.02E−02**	4.58E−02	4.72E−02
f_6	$\bar{f}$	2.15E−02	1.05E−02	1.19E−02	1.57E−02	1.59E−02	1.67E−02	1.69E−02	**7.94E−03**	1.98E−02	1.92E−02
	σ_f	1.90E−02	3.86E−03	1.13E−02	1.56E−02	1.49E−02	1.37E−02	1.49E−02	**1.89E−03**	1.80E−02	9.92E−03
f_7	$\bar{f}$	8.98E−03	3.22E−03	2.22E−03	1.88E−03	1.75E−03	2.07E−03	1.79E−03	**1.36E−03**	1.59E−03	1.64E−03
	σ_f	1.27E−02	2.65E−03	2.17E−03	1.48E−03	1.62E−03	2.26E−03	1.92E−03	**1.10E−03**	1.45E−03	1.50E−03
f_8	$\bar{f}$	−4.95E+03	−6.12E+03	−5.01E+04	−5.13E+04	−4.96E+03	−3.02E+03	−5.14E+04	**−5.58E+04**	−4.94E+03	−5.21E+03
	σ_f	4.36E+02	5.09E+02	4.15E+02	5.55E+02	5.08E+02	4.78E+02	4.24E+02	**4.10E+02**	4.85E+02	4.60E+02
f_9	$\bar{f}$	6.20E+01	2.91E+01	1.70E+01	1.57E+01	5.94E+00	5.21E+00	5.86E+00	**4.76E−01**	9.94E+00	2.02E+01
	σ_f	6.08E+01	5.32E+01	4.17E+01	4.27E+01	3.17E+01	2.77E+01	2.90E+01	**2.38E+00**	3.77E+01	5.22E+01
f_{10}	$\bar{f}$	8.70E−15	**7.16E−15**	7.99E−15	9.18E−15	9.41E−15	1.04E−14	1.19E−14	8.47E−15	1.07E−14	1.38E−14
	σ_f	5.39E−15	3.92E−15	3.61E−15	**2.53E−15**	2.57E−15	5.22E−15	6.00E−15	3.82E−15	6.64E−15	7.83E−15

(continued)

Table 8.2 (continued)

α_1		0.6	0.7	0.8	0.9	1	1.2	1.3	1.4	1.5	1.6
f_{11}	$\bar{f}$	**0.00E+00**	**0.00E+00**	**0.00E+00**	**0.00E+00**	**0.00E+00**	**0.00E+00**	**0.00E+00**	**0.00E+00**	**0.00E+00**	3.46E−05
	σ_f	**0.00E+00**	**0.00E+00**	**0.00E+00**	**0.00E+00**	**0.00E+00**	**0.00E+00**	**0.00E+00**	**0.00E+00**	**0.00E+00**	1.32E−04
f_{12}	$\bar{f}$	8.76E−02	8.32E−02	8.42E−02	7.82E−02	**7.70E−02**	8.73E−02	9.45E−02	9.59E−01	3.84E+00	6.49E+02
	σ_f	2.74E−02	3.35E−02	4.73E−02	2.30E−02	**1.80E−02**	2.78E−02	2.15E−02	3.26E+00	8.79E+00	2.65E+03
f_{13}	$\bar{f}$	**2.88E−01**	3.30E−01	3.69E−01	3.77E−01	3.99E−01	3.72E−01	4.27E−01	3.91E−01	2.44E+01	1.83E+06
	σ_f	**1.42E−01**	1.03E−01	1.63E−01	1.57E−01	2.20E−01	1.16E−01	3.78E−01	1.76E−01	9.67E+01	1.00E+07
f_{14}	$\bar{f}$	−8.43E+02	−8.33E+02	−8.31E+02	−8.29E+02	−8.43E+02	−8.97E+02	**−8.98E+02**	−8.90E+02	−8.86E+02	−8.84E+02
	σ_f	1.14E+01	9.37E+00	**1.06E+01**	1.12E+01	1.68E+01	2.54E+01	2.19E+01	1.80E+01	2.13E+01	2.48E+01

Bold elements represent the best values

In the second stage, the performance of the Fuzzy algorithm is evaluated considering different values for α_2. In the experiment, the values of α_2 are varied from 0.01 to 0.1 whereas the values of α_1 and γ remain fixed at 1.4 and 0.005, respectively. The statistical results obtained by the fuzzy algorithm using different values of α_2 are presented in Table 8.3. From Table 8.3, it is clear that our fuzzy optimization algorithm with $\alpha_2 = 0.05$ outperforms the other parameter configurations. Under this configuration, the algorithm obtains the best results in 8 of the 14 functions. However, if another parameter set is used, it results in a bad performance.

Finally, in the third stage, the performance of the Fuzzy algorithm is evaluated considering different values for γ. In the simulation, the values of γ are varied from 0.001 to 0.01 whereas the values of α_1 and α_2 remain fixed at 1.4 and 0.05, respectively. Table 8.4 summarizes the results of this experiment. From the information provided by Table 8.4, it can be seen that the fuzzy algorithm with $\gamma = 0.005$ obtains the best performance on functions $f_1, f_2, f_3, f_4, f_6, f_7, f_{10}, f_{12}$ and f_{13}. However, when the parameter γ takes any other value, the performance of the algorithm is inconsistent. Under this configuration, the algorithm presents the best possible performance, since it obtains the best indexes in 10 out of 14 functions.

In general, the experimental results shown in Tables 8.2, 8.3 and 8.4 suggest that a proper combination of the parameter values can improve the performance of the Fuzzy method and the quality of solutions. In this experiment we can conclude that the best parameter set is composed by the following values: $\alpha_1 = 1.4$, $\alpha_2 = 0.05$ and $\gamma = 0.005$.

Once the parameters α_1, α_2 and γ have been experimentally set, it is possible to analyze their influence in the optimization process. In the search strategy, integrated in the fuzzy system, α_1 modifies the attraction that a promising individual experiments with regard to the best current element in the population. This action aims to improve the solution quality of the individual, considering that the unexplored region between the promising solution and the best element could contain a better solution. On the other hand, α_2 adjusts the repulsion to which a low quality individual is undergone. This operation intends to enhance the quality of the bad candidate solution through a movement in opposite direction of the best current element. This repulsion is considered, since there is evidence that the unexplored section between the low quality solution and the best current element does not enclose promising solutions. Finally, γ defines the neighborhood around a promising solution, from which a local search operation is conducted. The objective of this process is to refine the quality of each solution that initially maintains an acceptable fitness value.

Considering their magnitude, the values of $\alpha_1 = 1.4$, $\alpha_2 = 0.05$ and $\gamma = 0.005$ indicate that the attraction procedure is the most important operation in the optimization strategy. This fact confirms that the attraction process represents the most prolific operation in the fuzzy strategy, since it searches new solutions in the direction where high fitness values are expected. According to its importance, the repulsion operation holds the second position. Repulsion produces significant small modifications of candidate solutions in comparison to the attraction process. This result indicates that the repulsion process involves an exploration with a higher

Table 8.3 Experimental results obtained by the fuzzy algorithm using different values of α_2

α_2		0.01	0.02	0.03	0.04	0.05	0.06	0.07	0.08	0.09	0.1
f_1	$\bar{f}$	1.01E−39	3.25E−28	3.39E−28	2.35E−28	**5.18E−49**	1.48E−28	1.36E−28	2.61E−28	2.15E−28	2.29E−28
	σ_f	5.54E−39	9.09E−28	1.02E−27	7.44E−28	**2.34E−48**	5.19E−28	4.95E−28	8.03E−28	6.36E−28	6.94E−28
f_2	$\bar{f}$	6.25E−16	3.61E−16	1.67E−22	2.66E−20	7.16E−22	1.15E−19	7.85E−16	3.29E−17	4.82E−16	**6.57E−30**
	σ_f	2.85E−15	1.97E−15	9.11E−22	1.45E−19	3.89E−21	5.41E−19	2.97E−15	1.53E−16	2.47E−15	**3.59E−29**
f_3	$\bar{f}$	4.96E−24	9.65E−27	9.29E−26	2.22E−25	**1.97E−27**	2.52E−23	2.81E−21	4.94E−22	7.99E−23	6.69E−21
	σ_f	2.66E−23	3.49E−26	4.33E−25	1.12E−24	**2.02E−27**	1.37E−22	1.53E−20	2.64E−21	2.25E−22	2.13E−20
f_4	$\bar{f}$	1.96E−15	1.20E−15	3.44E−17	5.99E−18	**3.08E−29**	1.92E−20	4.91E−26	6.13E−19	3.98E−16	1.42E−28
	σ_f	5.12E−15	4.45E−15	1.33E−16	3.28E−17	**1.69E−28**	9.01E−20	2.69E−25	3.31E−18	2.18E−15	7.76E−28
f_5	$\bar{f}$	3.85E+01	3.85E+01	2.55E+01	1.55E+01	**1.99E−04**	1.85E−01	4.85E−02	1.23E+01	1.35E+01	2.85E+01
	σ_f	4.45E−02	4.42E−02	4.38E−02	4.52E−02	**3.74E−02**	5.02E−02	3.45E−02	4.12E−02	5.02E−02	4.04E−02
f_6	$\bar{f}$	1.55E−02	1.47E−02	2.11E−02	2.15E−02	**1.07E−03**	1.73E−02	1.94E−02	1.78E−02	2.35E−01	2.04E−02
	σ_f	9.87E−03	5.13E−03	2.10E−02	4.72E−03	**1.67E−02**	6.85E−03	1.90E−02	7.59E−03	1.17E+00	2.21E−02
f_7	$\bar{f}$	**1.03E−03**	1.56E−03	1.09E−03	1.42E−03	1.36E−03	2.17E−03	1.88E−03	2.12E−03	2.53E−03	2.55E−03
	σ_f	8.62E−04	1.60E−03	7.41E−04	**1.07E−03**	1.10E−03	1.22E−03	1.56E−03	2.59E−03	3.04E−03	2.10E−03
f_8	$\bar{f}$	−3.10E+03	−5.24E+03	−2.17E+03	−5.00E+03	−5.13E+03	−5.01E+03	**−6.29E+03**	−5.11E+03	−5.24E+03	−5.18E+03
	σ_f	5.08E+02	4.49E+02	4.65E+02	3.83E+02	4.77E+02	**3.68E+02**	5.35E+02	4.36E+02	5.03E+02	4.41E+02
f_9	$\bar{f}$	8.98E+00	3.41E−02	5.91E+00	**9.16E−02**	4.76E−01	8.28E+00	7.75E−02	4.07E+00	1.64E+01	5.83E+00
	σ_f	3.42E+01	**1.87E−01**	3.14E+01	2.81E−01	2.38E+00	3.15E+01	2.61E−01	2.18E+01	5.02E+01	3.13E+01
f_{10}	$\bar{f}$	1.12E−14	1.01E−14	1.10E−14	1.17E−14	**8.47E−15**	9.30E−15	9.89E−15	1.21E−14	1.21E−14	1.20E−14
	σ_f	4.88E−15	3.82E−15	4.86E−15	7.26E−15	**3.44E−15**	4.71E−15	3.82E−15	6.19E−15	7.11E−15	6.03E−15

(continued)

Table 8.3 (continued)

α_2		0.01	0.02	0.03	0.04	0.05	0.06	0.07	0.08	0.09	0.1
f_{11}	$\bar{f}$	**0.00E+00**	**0.00E+00**	**0.00E+00**	**0.00E+00**	**0.00E+00**	**0.00E+00**	**0.00E+00**	**0.00E+00**	**0.00E+00**	**0.00E+00**
	σ_f	**0.00E+00**	**0.00E+00**	**0.00E+00**	**0.00E+00**	**0.00E+00**	**0.00E+00**	**0.00E+00**	**0.00E+00**	**0.00E+00**	**0.00E+00**
f_{12}	$\bar{f}$	1.03E+00	7.77E−01	**9.55E−02**	1.93E+00	9.59E−01	4.14E−01	1.17E+00	1.26E+00	1.60E+00	6.41E+00
	σ_f	3.80E+00	2.63E+00	**3.50E−02**	4.30E+00	3.26E+00	1.75E+00	3.56E+00	4.41E+00	4.70E+00	2.28E+01
f_{13}	$\bar{f}$	3.54E−01	4.08E−01	1.69E+00	2.38E+00	**3.51E−01**	4.20E−01	1.17E+00	1.12E+00	8.94E−01	1.34E+00
	σ_f	1.91E−01	2.18E−01	5.25E+00	7.65E+00	**1.76E−01**	1.69E−01	4.03E+00	3.38E+00	2.53E+00	4.99E+00
f_{14}	$\bar{f}$	−8.85E+02	−8.84E+02	−8.91E+02	−8.88E+02	−8.90E+02	−8.87E+02	−8.82E+02	−8.86E+02	**−8.94E+02**	−8.82E+02
	σ_f	2.41E+01	1.81E+01	2.25E+01	2.39E+01	1.80E+01	**1.47E+01**	2.37E+01	1.63E+01	2.07E+01	1.54E+01

Bold elements represent the best values

Table 8.4 Experimental results obtained by the fuzzy algorithm using different values of γ

γ		0.001	0.002	0.003	0.004	0.005	0.006	0.007	0.008	0.009	0.01
f_1	$\bar{f}$	3.73E−28	5.52E−29	4.79E−29	1.14E−28	**1.01E−39**	2.04E−28	1.92E−28	1.20E−28	8.28E−29	1.27E−28
	σ_f	8.01E−28	2.74E−28	2.23E−28	5.10E−28	**5.54E−39**	7.66E−28	7.31E−28	4.91E−28	4.46E−28	6.93E−28
f_2	$\bar{f}$	1.78E−16	5.62E−16	6.03E−16	9.19E−17	**5.26E−35**	4.20E−16	2.44E−22	7.16E−22	3.97E−17	2.41E−18
	σ_f	8.72E−16	2.14E−15	2.30E−15	4.21E−16	**2.88E−34**	1.63E−15	1.34E−21	3.89E−21	2.18E−16	1.30E−17
f_3	$\bar{f}$	1.19E−23	5.91E−25	1.31E−23	1.74E−24	**2.35E−25**	4.96E−24	6.38E−22	1.91E−23	3.41E−21	7.75E−13
	σ_f	6.49E−23	2.61E−24	6.52E−23	5.07E−24	**8.29E−25**	2.66E−23	2.64E−21	1.04E−22	1.85E−20	4.24E−12
f_4	$\bar{f}$	3.34E−26	9.36E−16	5.20E−16	5.23E−16	**6.13E−19**	7.80E−18	5.46E−16	6.14E−16	7.75E−21	4.90E−16
	σ_f	1.52E−25	3.79E−15	2.85E−15	2.86E−15	**3.31E−18**	4.27E−17	2.99E−15	2.37E−15	4.24E−20	2.56E−15
f_5	$\bar{f}$	2.75E+01	2.85E+01	2.97E+01	**3.85E−04**	1.45E−01	3.55E−01	8.35E−01	1.23E+00	2.78E+01	2.85E+01
	σ_f	4.67E−02	4.29E−02	4.47E−02	**2.85E−02**	4.38E−02	4.52E−02	4.31E−02	3.96E−02	4.18E−02	4.49E−02
f_6	$\bar{f}$	1.89E−02	1.93E−02	1.60E−02	1.85E−02	**1.54E−02**	1.57E−02	2.17E−02	2.06E−02	2.15E−02	1.92E−02
	σ_f	1.27E−02	1.50E−02	7.99E−03	1.14E−02	**5.55E−03**	8.61E−03	1.45E−02	1.91E−02	1.90E−02	1.34E−02
f_7	$\bar{f}$	1.98E−03	1.76E−03	1.49E−03	1.58E−03	**1.32E−03**	1.71E−03	1.61E−03	1.95E−03	2.30E−03	1.36E−03
	σ_f	2.02E−03	1.62E−03	2.01E−03	1.56E−03	**1.03E−03**	1.38E−03	1.92E−03	2.03E−03	2.79E−03	1.10E−03
f_8	$\bar{f}$	−5.11E+02	−5.05E+03	**−5.27E+04**	−5.19E+03	−5.13E+04	−4.98E+03	−5.05E+03	−4.12E+02	−5.11E+02	−4.98E+03
	σ_f	5.20E+02	4.42E+02	5.69E+02	4.20E+02	4.77E+02	4.66E+02	4.54E+02	5.31E+02	**3.24E+02**	5.47E+02
f_9	$\bar{f}$	**1.14E−14**	6.94E+00	4.72E+00	1.07E+01	4.76E−01	6.13E+00	8.69E+00	2.16E+01	7.27E−02	1.41E+01
	σ_f	**2.75E−14**	2.55E+01	2.56E+01	4.05E+01	2.38E+00	3.18E+01	3.15E+01	5.64E+01	2.81E−01	4.31E+01
f_{10}	$\bar{f}$	1.23E−14	8.70E−15	9.06E−15	1.13E−14	**1.04E−14**	1.05E−14	9.06E−15	1.26E−14	8.47E−15	8.82E−15
	σ_f	6.29E−15	3.29E−15	2.97E−15	6.79E−15	**2.97E−15**	6.06E−15	3.82E−15	6.47E−15	5.55E−15	3.58E−15

(continued)

Table 8.4 (continued)

γ		0.001	0.002	0.003	0.004	0.005	0.006	0.007	0.008	0.009	0.01
f_{11}	$\bar{f}$	**0.00E+00**	**0.00E+00**	**0.00E+00**	**0.00E+00**	**0.00E+00**	4.78E−05	**0.00E+00**	**0.00E+00**	**0.00E+00**	**0.00E+00**
	σ_f	**0.00E+00**	**0.00E+00**	**0.00E+00**	**0.00E+00**	**0.00E+00**	1.85E−04	**0.00E+00**	**0.00E+00**	**0.00E+00**	**0.00E+00**
f_{12}	$\bar{f}$	2.11E+00	9.59E−01	9.95E−01	7.95E−01	**9.38E−02**	8.96E−01	5.60E−01	1.08E+00	1.32E−01	6.74E−01
	σ_f	7.38E+00	3.26E+00	3.37E+00	3.83E+00	**1.99E−02**	3.17E+00	2.48E+00	3.73E+00	1.76E−01	3.12E+00
f_{13}	$\bar{f}$	3.98E−01	4.28E−01	4.12E−01	3.90E−01	**3.85E−01**	1.06E+00	3.91E−01	4.08E−01	1.14E+00	3.88E−01
	σ_f	2.68E−01	2.24E−01	3.49E−01	1.90E−01	**1.51E−01**	3.82E+00	1.76E−01	2.24E−01	4.12E+00	2.08E−01
f_{14}	$\bar{f}$	−8.86E+02	−8.82E+02	−8.91E+02	−8.91E+02	−8.90E+02	−8.93E+02	−8.93E+02	**−8.97E+02**	−8.89E+02	−8.89E+02
	σ_f	2.29E+01	2.03E+01	2.22E+01	2.03E+01	**1.80E+01**	2.72E+01	1.91E+01	2.24E+01	1.97E+01	1.98E+01

Bold elements represent the best values

uncertainty compared with the attraction movement. This uncertainty is a consequence of the lack of knowledge, if the opposite movement may reach a position with a better fitness value. The only available evidence is that in direction of the attraction movement, it is not possible to find promising solutions. Finally, the small value of γ induces a minor vibration for each acceptable candidate solution, in order to refine its quality in terms of fitness value.

8.5.2 Comparison with Other Optimization Approaches

In this subsection, the Fuzzy method is evaluated in comparison with other popular optimization algorithms based on evolutionary principles. In the experiments, we have applied the fuzzy optimization algorithm to the 19 functions from appendix, and the results are compared to those produced by the Harmony Search (HS) method [14], the Bat (BAT) algorithm [15], the Differential Evolution (DE) [19], the Particle Swarm Optimization (PSO) method [12], the Artificial Bee Colony (ABC) algorithm [13] and the Co-variance Matrix Adaptation Evolution Strategies (CMA-ES) [49]. These are considered the most popular metaheuristic algorithms currently in use [50]. In the experiments, the population size N has been configured to 50 individuals. The operation of the benchmark functions is conducted in 50 and 100 dimensions. In order to eliminate the random effect, each function is tested for 30 independent runs. In the comparison, a fixed number FN of function evaluations has been considered as a stop criterion. Therefore, each execution of a test function consists of $FN = 10^4 \cdot n$ function evaluations (where n represents the number of dimensions). This stop criterion has been decided to keep compatibility with similar works published in the literature [51–54].

For the comparison, all methods have been configured with the parameters, which according to their reported references reach their best performance. Such configurations are described as follows:

1. **HS** [14]: $HCMR = 0.7$ and $PArate = 0.3$.
2. **BAT** [15]: Loudness ($A = 2$), Pulse Rate ($r = 0.9$), Frequency minimum ($Q_{\min} = 0$) and Frequency maximum ($Q_{\min} = 1$).
3. **DE** [19]: $CR = 0.5$ and $F = 0.2$.
4. **PSO** [12]: $c_1 = 2$ and $c_2 = 2$; the weight factor decreases linearly from 0.9 to 0.2.
5. **ABC** [13]: $limit = 50$.
6. **CMA-ES** [47]: The source code has been obtained from the original author [55]. In the experiments, some minor changes have been applied to adapt CMA-ES to our test functions, but the main body is unaltered.
7. **FUZZY**: $\alpha_1 = 1.4$, $\alpha_2 = 0.05$ and $\gamma = 0.005$.

Several tests have been conducted for comparing the performance of the fuzzy algorithm. The experiments have been divided in Unimodal functions (Table 8.17), Multimodal functions (Table 8.18) and Hybrid functions (Table 8.19).

8.5.2.1 Unimodal Test Functions

In this test, the performance of our fuzzy algorithm is compared with HS, BAT, DE, PSO, CMA-ES and ABC, considering functions with only one optimum. Such functions are represented by functions f_1 to f_7 in Table 8.17. In the test, all functions have been operated in 50 dimensions ($n = 50$). The experimental results obtained from 30 independent executions are presented in Table 8.5. They report the averaged best fitness values ($\bar{f}$) and the standard deviations (σ_f) obtained in the runs. We have also included the best (f_{Best}) and the worst (f_{Worst}) fitness values obtained during the total number of executions. The best entries in Table 8.5 are highlighted in boldface. From Table 8.5, according to the averaged best fitness value ($\bar{f}$) index, we can conclude that the Fuzzy method performs better than the other algorithms in functions f_1, f_3, f_4 and f_7. In the case of functions f_2, f_5 and f_6, the CMA-ES algorithm obtains the best results. By contrast, the rest of the algorithms presents different levels of accuracy, with ABC being the most consistent. These results indicate that the Fuzzy approach provides better performance than HS, BAT, DE, PSO and ABC for all functions except for the CMA-ES which delivers similar results to those produced by the Fuzzy approach. By analyzing the standard deviation (σ_f) index in Table 8.5, it becomes clear that the metaheuristic method which presents the best results also normally obtains the smallest deviations.

To statistically analyze the results of Table 8.5, a non-parametric test known as the Wilcoxon analysis [56, 57] has been conducted. It allows us to evaluate the differences between two related methods. The test is performed for the 5% (0.05) significance level over the "averaged best fitness values ($\bar{f}$)" data. Table 8.6 reports the p-values generated by Wilcoxon analysis for the pair-wise comparison of the algorithms. For the analysis, five groups are produced: FUZZY versus HS, FUZZY versus BAT, FUZZY versus DE, FUZZY versus PSO, FUZZY versus CMA-ES and FUZZY versus ABC. In the Wilcoxon analysis, it is considered a null hypothesis that there is no notable difference between the two methods. On the other hand, it is admitted as an alternative hypothesis that there is an important difference between the two approaches. In order to facilitate the analysis of Table 8.6, the symbols ▲, ▼, and ► are adopted. ▲ indicates that the Fuzzy method performs significantly better than the tested algorithm on the specified function. ▼ symbolizes that the Fuzzy algorithm performs worse than the tested algorithm, and ► means that the Wilcoxon rank sum test cannot distinguish between the simulation results of the fuzzy optimizer and the tested algorithm. The number of cases that fall in these situations are shown at the bottom of the table.

After an analysis of Table 8.6, it is evident that all p-values in the FUZZY versus HS, FUZZY versus BAT, FUZZY versus DE, FUZZY versus PSO and FUZZY

Table 8.5 Minimization results of unimodal functions of Table 8.17 with n = 50

		HS	BAT	DE	PSO	CMA-ES	ABC	FUZZY
f_1	$\bar{f}$	87035.2235	121388.0212	61.1848761	4.39E+03	1.34E−11	3.09E−06	**2.30E−29**
	σ_f	5262.26532	6933.129294	163.555175	1261.19173	5.1938E−12	3.4433E−06	**1.17237E−28**
	f_{Best}	76937.413	108807.878	0.03702664	1.65E+03	5.69E−12	2.47E−07	**5.17E−114**
	f_{Worst}	95804.9747	138224.1125	878.436103	7.37E+03	2.55E−11	1.72E−05	**6.42E−28**
f_2	$\bar{f}$	1.3739E+14	4.31636E+17	0.04057031	4.54E+01	**9.92E−06**	1.39E−03	4.15E−04
	σ_f	3.188E+14	1.53734E+18	0.09738928	16.386199	**2.5473E−06**	0.00071159	0.00227186
	f_{Best}	1.0389E+10	1633259021	4.03E−12	2.61E+01	5.87E−06	5.62E−04	**7.20E−59**
	f_{Worst}	1.64E+15	7.60E+18	0.45348954	9.75E+01	**1.44E−05**	2.98E−03	0.01244379
f_3	$\bar{f}$	130472.801	297342.4211	55982.8182	1.57E+04	2.89E−03	4.14E+04	**1.93E−05**
	σ_f	11639.2864	99049.83213	9234.85975	9734.92204	0.00164804	4785.18216	**4.2843E−05**
	f_{Best}	104514.012	164628.01	36105.5799	4.23E+03	9.88E−04	2.85E+04	**1.66E−10**
	f_{Worst}	147659.604	563910.1737	70938.4205	4.96E+04	8.88E−03	4.84E+04	**0.00018991**
f_4	$\bar{f}$	80.1841708	90.17564768	25.8134455	2.32E+01	3.96E−04	7.35E+01	**3.37E−16**
	σ_f	2.55950002	1.862675447	6.30765469	3.51409694	8.2083E−05	3.60905231	**1.8484E−15**
	f_{Best}	73.2799506	86.11297617	15.7894785	1.73E+01	2.57E−04	6.55E+01	**7.52E−70**
	f_{Worst}	83.8375161	92.78058061	38.8210447	3.06E+01	5.65E−04	7.90E+01	**1.01E−14**
f_5	$\bar{f}$	1024.70257	276.2438329	52.5359064	6.04E+02	**3.51E−05**	4.53E+01	4.85E−04
	σ_f	100.932656	45.12095642	7.69858817	198.334321	0.49723274	1.13628434	**0.0389642**
	f_{Best}	783.653134	211.6001157	47.1421071	289.29993	**1.21E−09**	42.1783081	3.30E−09
	f_{Worst}	1211.08532	399.1608511	75.1362468	1126.38574	**3.0654249**	47.7422282	4.6323356

(continued)

Table 8.5 (continued)

		HS	BAT	DE	PSO	CMA-ES	ABC	FUZZY
f_6	$\bar{f}$	88027.4244	119670.6412	43.5155273	4.51E+03	**1.42E−11**	4.15E−06	2.18E−07
	σ_f	5783.21576	6818.723503	80.4217558	2036.72193	**5.5321E−12**	8.5588E−06	0.84607249
	f_{Best}	77394.5062	105958.6224	0.01832758	1705.47866	**5.88E−12**	6.00E−07	1.18513127
	f_{Worst}	97765.4819	130549.7364	306.098587	13230.6439	**2.85E−11**	4.79E−05	5.18913374
f_7	$\bar{f}$	197.476174	116.8196698	0.08164158	4.43E+01	2.82E−02	6.86E−01	**3.43E−04**
	σ_f	28.808573	16.46542385	0.12240289	17.8200508	0.00499868	0.14547266	**0.00447976**
	f_{Best}	116.483527	87.64501186	0.01387586	15.7697307	0.0201694	0.41798576	**0.00018152**
	f_{Worst}	263.233333	156.0245904	0.65353574	85.526355	0.03888318	0.8957427	**0.02057915**

Bold elements represent the best values

Table 8.6 p-values produced by Wilcoxon test comparing FUZZY versus HS, FUZZY versus BAT, FUZZY versus DE, FUZZY versus PSO, FUZZY versus CMA-ES and FUZZY versus ABC over the "averaged best fitness values" from Table 8.5

Wilcoxon test for unimodal functions of Table 8.17 with $n = 50$						
FUZZY versus	HS	BAT	DE	PSO	CMA-ES	ABC
f_1	5.0176E−07▲	9.4988E−08▲	7.3648E−07▲	7.8952E−05▲	7.234E−03▲	5.1480E−04▲
f_2	4.0553E−07▲	2.4620E−08▲	2.0793E−03▲	2.0182E−05▲	0.0937►	3.0415E−03▲
f_3	2.0189E−08▲	3.7451E−08▲	1.0492E−07▲	4.1590E−05▲	0.0829►	2.7612E−06▲
f_4	3.5470E−07▲	2.1490E−08▲	3.4081E−06▲	2.0121E−06▲	8.143E−03▲	4.1680E−07▲
f_5	1.0795E−08▲	4.0479E−09▲	2.0354E−07▲	8.1350E−09▲	0.1264►	1.2541E−07▲
f_6	6.1769E−07▲	6.5480E−08▲	4.5972E−06▲	2.1594E−07▲	0.0741►	2.1548E−03▲
f_7	4.3617E−07▲	1.9235E−08▲	2.8070E−04▲	5.4890E−06▲	0.1031►	1.0430E−03▲
▲	7	7	7	7	2	7
▼	0	0	0	0	0	0
►	0	0	0	0	5	0

versus ABC columns are less than 0.05 (5% significance level) which is a strong evidence against the null hypothesis and indicates that the Fuzzy method performs better (▲) than the HS, BAT, DE, PSO and ABC algorithms. This data is statistically significant and shows that it has not occurred by coincidence (i.e. due to the normal noise contained in the process). In the case of the comparison between FUZZY and CMA-ES, the FUZZY method maintains a better (▲) performance in functions f_1 and f_4. In functions f_2, f_3, f_5, f_6 and f_7 the CMA-ES presents a similar performance to the FUZZY method. This fact can be seen from the column FUZZY versus CMA-ES, where the p-values of functions f_2, f_3, f_5, f_6 and f_7 are higher than 0.05 (►). These results reveal that there is no statistical difference in terms of precision between FUZZY and CMA-ES, when they are applied to the aforementioned functions. In general, the results of the Wilcoxon analysis demonstrates that the Fuzzy algorithm performs better than most of the other methods.

In addition to the experiments in 50 dimensions, we have also conducted a set of simulations on 100 dimensions to test the scalability of the fuzzy method. In the analysis, we also employed all the compared algorithms in this test. The simulation results are presented in Tables 8.7 and 8.8, which report the data produced during the 30 executions and the Wilcoxon analysis, respectively. According to the averaged best fitness value ($\bar{f}$) index from Table 8.7, the Fuzzy method performs better than the other algorithms in functions f_1, f_2, f_3, f_4 and f_7. In the case of functions f_5 and f_6, the CMA-ES algorithm obtains the best results. On the other

Table 8.7 Minimization results of unimodal functions of Table 8.17 with $n = 100$

		HS	BAT	DE	PSO	CMA-ES	ABC	FUZZY
f_1	$\bar{f}$	2.19E+05	2.63E+05	3.89E+02	1.43E+04	1.32E−05	2.45E−02	**1.89E−16**
	σ_f	10311.7753	14201.563	323.607739	2920.78144	3.2095E−06	0.02575058	**1.0358E−15**
	f_{Best}	1.74E+05	2.30E+05	1.26E+01	9.64E+03	8.00E−06	4.46E−03	**6.43E−68**
	f_{Worst}	2.30E+05	2.89E+05	1.38E+03	2.09E+04	2.04E−05	1.26E−01	**5.67E−15**
f_2	$\bar{f}$	7.24E+37	1.31E+45	5.73E−01	1.51E+02	1.26E−02	9.28E−02	**1.89E−08**
	σ_f	2.587E+38	6.9056E+45	0.57775559	41.1235147	0.00287131	0.02545392	**1.03445E−07**
	f_{Best}	5.45E+31	3.36E+34	2.94E−02	9.05E+01	8.91E−03	5.40E−02	**4.84E−46**
	f_{Worst}	1.35E+39	3.79E+46	2.56E+00	2.52E+02	2.38E−02	0.18353771	**5.67E−07**
f_3	$\bar{f}$	4.98E+05	1.15E+06	2.84E+05	7.90E+04	8.76E−04	1.84E+05	**2.07E−08**
	σ_f	58467.2769	312595.437	27132.9955	34174.7164	0.7743521	20108.5821	**0.54899784**
	f_{Best}	3.37E+05	4.69E+05	2.31E+05	3.71E+04	8.76E−04	1.27E+05	**3.94E−09**
	f_{Worst}	616974.994	1942095.52	356515.579	160139.954	145362.58	219982.397	**2.25159901**
f_4	$\bar{f}$	9.01E+01	9.46E+01	4.05E+01	2.81E+01	2.16E−06	9.04E+01	**2.63E−15**
	σ_f	1.11508888	0.85959226	6.84190892	3.2581793	0.03982112	1.8297347	**6.0638E−15**
	f_{Best}	8.69E+01	9.29E+01	2.70E+01	2.15E+01	1.39E−08	8.37E+01	**1.24E−67**
	f_{Worst}	9.16E+01	9.62E+01	5.55E+01	3.40E+01	2.91E−01	9.30E+01	**2.02E−14**
f_5	$\bar{f}$	2.70E+03	1.15E+03	1.29E+02	3.95E+03	**9.09E−04**	1.04E+02	9.86E−04
	σ_f	540.831375	88.3793364	18.7799545	641.937017	1.45575683	6.27511831	**0.14584341**
	f_{Best}	**0**	980.500576	101.873611	2571.71519	**1.32E−05**	96.9508931	2.43E−06
	f_{Worst}	3247.18778	1308.54456	167.973469	5316.76433	**94.8276069**	121.37168	99.203712

(continued)

Table 8.7 (continued)

		HS	BAT	DE	PSO	CMA-ES	ABC	FUZZY
f_6	$\bar{f}$	2.21E+05	2.64E+05	3.70E+02	1.45E+04	**1.51E−05**	1.92E−02	2.02E−05
	σ_f	9381.48118	18216.585	280.553973	2798.89068	**2.2704E−06**	0.01816388	2.87413087
	f_{Best}	198863.662	226296.005	16.8023385	9243.44125	**1.10E−05**	0.0025079	0.15E−05
	f_{Worst}	235307.769	288557.483	1278.66865	20954.2719	**1.94E−05**	0.09120152	23.7436855
f_7	$\bar{f}$	1.28E+03	4.67E+02	7.35E−01	4.85E+02	7.03E−02	2.20E+00	**4.51E−03**
	σ_f	100.811769	54.7947564	0.58830651	125.201537	0.01043989	0.36193003	**0.00535871**
	f_{Best}	975.480173	351.38694	0.08400938	233.702775	0.04689927	1.43278953	**3.59E−05**
	f_{Worst}	1424.35137	609.120842	2.38315757	806.350617	0.08989015	2.98777544	**0.0213576**

Bold elements represent the best values

Table 8.8 p-values produced by Wilcoxon test comparing FUZZY versus HS, FUZZY versus BAT, FUZZY versus DE, FUZZY versus PSO, FUZZY versus CMA-ES and FUZZY versus ABC over the "averaged best fitness values" from Table 8.7

Wilcoxon test for unimodal functions of Table 8.17 with n = 100						
FUZZY versus	HS	BAT	DE	PSO	CMA-ES	ABC
f_1	3.0150E−08▲	8.0794E−08▲	6.4901E−06▲	2.0146E−07▲	7.9460E−04▲	4.0168E−05▲
f_2	5.3480E−11▲	1.1302E−11▲	7.4985E−05▲	4.7001E−06▲	2.4920E−04▲	8.2940E−04▲
f_3	6.0145E−07▲	7.0651E−08▲	4.6782E−07▲	8.4670E−06▲	0.0743►	2.3014E−07▲
f_4	1.4920E−07▲	3.7912E−07▲	2.0142E−06▲	1.4972E−06▲	7.4682E−04▲	2.1966E−07▲
f_5	8.7942E−06▲	5.4972E−06▲	9.4662E−05▲	2.1580E−07▲	0.1851►	4.7223E−05▲
f_6	2.7301E−07▲	4.7920E−07▲	8.0493E−05▲	5.7942E−06▲	0.2451►	1.4901E−04▲
f_7	1.0458E−07▲	5.4201E−06▲	2.0051E−04▲	7.6190E−06▲	0.0851►	4.610E−05▲
▲	7	7	7	7	3	7
▼	0	0	0	0	0	0
►	0	0	0	0	4	0

hand, the rest of the algorithms present different levels of accuracy. After analyzing Table 8.7, it is clear that the fuzzy method presents slightly better results than CMA-ES in 100 dimensions. From Table 8.8, it is evident that all p-values in the FUZZY versus HS, FUZZY versus BAT, FUZZY versus DE, FUZZY versus PSO and FUZZY versus ABC columns are less than 0.05, which indicates that the Fuzzy method performs better than the HS, BAT, DE, PSO and ABC algorithms. In the case of FUZZY versus CMA-ES, the FUZZY method maintains a better performance in functions f_1, f_2 and f_4. In functions f_3, f_5, f_6 and f_7 the CMA-ES presents a similar performance to the FUZZY method. This experiment shows that the more dimensions there are, the worse the performance of the CMA-ES is.

8.5.2.2 Multimodal Test Functions

Contrary to unimodal functions, multimodal functions include many local optima. For this cause, they are, in general, more complicated to optimize. In this test the performance of our algorithm is compared with HS, BAT, DE, PSO, CMA-ES and ABC regarding multimodal functions. Multimodal functions are represented by functions from f_8 to f_{14} in Table 8.18, where the number of local minima increases exponentially as the dimension of the function increases. Under such conditions, the experiment reflects the ability of each algorithm to find the global optimum in the presence of numerous local optima. In the simulations, the functions are operated in 50 dimensions ($n = 50$). The results, averaged over 30 executions, are reported in Table 8.9 in terms of the best fitness values ($\bar{f}$) and the standard deviations (σ_f). The best results are highlighted in boldface. Likewise, p-values of the Wilcoxon test of 30 independent repetitions are exhibited in Table 8.10. In the case of f_8, f_{10}, f_{11} and f_{14}, the fuzzy method presents a better performance than HS, BAT, DE, PSO, CMA-ES and ABC. For functions f_{12} and f_{13}, the fuzzy approach exhibits a worse performance compared to CMA-ES. Additionally, in the case of function f_9 the Fuzzy method and ABC maintain the best performance compared to HS, BAT, DE, PSO and CMA-ES. The rest of the algorithms present different levels of accuracy, with ABC being the most consistent. In particular, this test yields a large difference in performance, which is directly related to a better trade-off between exploration and exploitation produced by the formulated rules of the fuzzy method.

The results of the Wilcoxon analysis, presented in Table 8.10, statistically demonstrate that the Fuzzy algorithm performs better than HS, DE, BAT, DE and PSO in all test functions (f_8–f_{14}). In the case of the comparison between FUZZY and CMA-ES, the FUZZY method maintains a better (▲) performance in functions f_8, f_9, f_{10}, f_{11} and f_{14}. On the other hand, in functions f_{12} and f_{13} the FUZZY method presents worse results (▼) than the CMA-ES algorithm. However, according to Table 8.10, the FUZZY approach obtains a better performance than ABC in all cases except for function f_9, where there is no difference in results between the two.

In addition to the 50-dimension benchmark function tests, we also performed a series of simulations with 100 dimensions by using the same set of functions in Table 8.18. The results are presented in Tables 8.11 and 8.12, which report the data produced during the 30 executions and the Wilcoxon analysis, respectively. In Table 8.11, it can be seen that the Fuzzy method performs better than HS, BAT, DE, PSO, CMA-ES and ABC for functions f_8, f_9, f_{10}, f_{11} and f_{13}. On the other hand, the CMA-ES maintains better results than HS, BAT, DE, PSO, ABC and the fuzzy optimizer for function f_{12}. Likewise, the DE method obtains better indexes than the other algorithms for function f_{14}. From the Wilcoxon analysis shown in Table 8.12, the results indicate that the Fuzzy method performs better than the HS, BAT, DE, PSO and ABC algorithms. In the case of FUZZY versus CMA-ES, the FUZZY method maintains a better performance in all test functions except in problem f_{12}, where the CMA-ES produces better results than the FUZZY method. This experiment also shows that the more dimensions there are, the worse the performance of the CMA-ES is.

8.5.2.3 Hybrid Test Functions

In this test, hybrid functions are employed to test the optimization performance of the Fuzzy approach. Hybrid functions, shown in Table 8.19, are multimodal functions with complex behaviors, since they are built from different multimodal single functions. A detailed implementation of the hybrid functions can be found in

Table 8.9 Minimization results of multimodal functions of Table 8.18 with n = 50

		HS	BAT	DE	PSO	CMA-ES	ABC	FUZZY
f_8	$\bar{f}$	−5415.83905	−3270.254967	−20232.0393	−1.00E+04	−6.22E+03	−1.94E+04	**−2.69E+05**
	σ_f	**318.326084**	474.2974644	799.670519	1139.72504	577.603827	328.074723	338131.298
	f_{Best}	−6183.25306	−4253.007751	−20830.6009	−12207.27	−7910.14987	−20460.0202	**−1602802.18**
	f_{Worst}	−4937.01673	−2560.016959	−16849.3789	−7417.20923	−5277.36121	−18598.5718	**−51860.8124**
f_9	$\bar{f}$	637.314967	370.181231	94.9321639	2.78E+02	9.12E−03	**6.43E−06**	1.03E−06
	σ_f	25.9077403	31.0789956	23.8913991	38.1965572	8.314267	**2.34825349**	1.6781696
	f_{Best}	581.055495	321.398632	49.5787092	2.09E+02	6.91E−04	1.99899276	**0**
	f_{Worst}	681.505155	450.919651	143.664199	375.979507	342.621828	**11.2277505**	310.43912
f_{10}	$\bar{f}$	20.2950743	19.2995398	1.04072229	1.21E+01	8.74E−07	1.69E−02	**1.40E−14**
	σ_f	0.09866263	0.11146929	0.69779278	1.03376556	1.4921E−07	0.01136192	**4.489E−15**
	f_{Best}	19.9003135	18.9741996	0.0040944	9.35E+00	5.62E−07	2.92E−03	**7.99E−15**
	f_{Worst}	20.472367	19.576839	2.78934934	1.38E+01	1.18E−06	5.30E−02	**2.22E−14**
f_{11}	$\bar{f}$	786.564993	1072.40695	0.98725915	4.05E+01	9.87E−10	6.23E−03	**0.00E+00**
	σ_f	49.0195978	70.0220465	0.62998733	12.8397453	4.8278E−10	0.01154936	**0**
	f_{Best}	658.158623	926.062051	0.00107768	14.1594978	2.92E−10	0.01503879	**0**
	f_{Worst}	860.983823	1186.93137	2.4153938	64.7187195	2.32E−09	0.053391	**0**
f_{12}	$\bar{f}$	557399404	1029876322	1309.87126	4.08E+01	**2.58E−12**	3.67E−07	1.91E−08
	σ_f	68320767.6	150067294	4319.40539	27.0146375	**1.0706E−12**	4.1807E−07	0.63138873
	f_{Best}	444164964	763229039	0.20366113	16.607083	**9.05E−13**	2.14E−08	1.95E−08
	f_{Worst}	700961313	1277767934	17508.9826	136.891908	**5.63E−12**	1.84E−06	19.8206283

(continued)

Table 8.9 (continued)

		HS	BAT	DE	PSO	CMA-ES	ABC	FUZZY
f_{13}	$\bar{f}$	1163989772	1982187734	29551.4297	1.41E+05	**5.03E−11**	5.98E−06	9.06E−09
	σ_f	123421334	291495991	137415.981	186740.994	**2.7928E−11**	7.5042E−06	0.0439066
	f_{Best}	898858903	1250159582	4.37613356	1594.63864	**1.37E−11**	5.06E−07	6.91E−10
	f_{Worst}	1453640537	2558128052	754699.3	735426.524	**1.38E−10**	3.48E−05	60.2604938
f_{14}	$\bar{f}$	−958.679663	−1066.80779	−1937.10075	−1.40E+03	−1.84E+03	−1.32E+03	**−1.96E+03**
	σ_f	40.8885038	61.5793102	13.199329	77.8992722	41.6063335	29.4904745	**0.06633944**
	f_{Best}	−1060.29598	−1213.46466	−1958.29881	−1527.69665	−1915.89813	−1395.05118	**−1958.30818**
	f_{Worst}	−893.619964	−988.87492	−1915.09727	−1275.90292	−1732.12078	−1262.8218	**−1958.04305**

Bold elements represent the best values

Table 8.10 p-values produced by Wilcoxon test comparing FUZZY versus HS, FUZZY versus BAT, FUZZY versus DE, FUZZY versus PSO, FUZZY versus CMA-ES and FUZZY versus ABC over the "averaged best fitness values" values from Table 8.9

Wilcoxon test for Multimodal functions of Table 8.18 with n = 50						
FUZZY versus	HS	BAT	DE	PSO	CMA-ES	ABC
f_8	7.1350E−05▲	5.4032E−05▲	3.4760E−05▲	1.1452E−05▲	4.7201E−06▲	2.4993E−05▲
f_9	4.1640E−05▲	2.4886E−05▲	6.1472E−04▲	2.1235E−05▲	4.2910E−04▲	0.0783►
f_{10}	3.1425E−05▲	3.0183E−05▲	7.4920E−04▲	1.4982E−05▲	3.1157E−04▲	9.4872E−04▲
f_{11}	5.4971E−08▲	9.3345E−09▲	7.1350E−04▲	5.3791E−06▲	8.4973E−03▲	6.1540E−03▲
f_{12}	6.4821E−11▲	8.4038E−11▲	4.6840E−08▲	5.2920E−06▲	7.2365E−04▼	4.0312E−03▲
f_{13}	7.9824E−11▲	9.7930E−11▲	4.1622E−10▲	7.4682E−11▲	9.4003E−04▼	4.5513E−05▲
f_{14}	7.1352E−06▲	4.5821E−07▲	5.7920E−04▲	8.1641E−05▲	9.6401E−04▲	5.6820E−05▲
▲	7	7	7	7	5	6
▼	0	0	0	0	2	0
►	0	0	0	0	0	1

[46]. In the experiments, the performance of our fuzzy algorithm is compared with HS, BAT, DE, PSO, CMA-ES and ABC, considering functions f_{15} to f_{19}.

In the first test, all functions have been operated in 50 dimensions (n = 50). The experimental results obtained from 30 independent executions are presented in Tables 8.13 and 8.14. In Table 8.13, the indexes $\bar{f}$, σ_f, f_{Best} and f_{Worst}, obtained during the total number of executions, are reported. Furthermore, Table 8.14 presents the statistical Wilcoxon analysis of the averaged best fitness values $\bar{f}$ from Table 8.13.

According to Table 8.13, the Fuzzy approach maintains a superior performance in comparison to most of the other methods. In the case of f_{15}, f_{16} and f_{18}, the fuzzy method performs better than HS, BAT, DE, PSO, CMA-ES and ABC. For function f_{19}, the fuzzy approach presents a worst performance than CMA-ES or ABC. However, in functions f_{16} and f_{18}, the Fuzzy method and ABC maintain a better performance than HS, BAT, DE, PSO and CMA-ES. For function f_{17} the FUZZY method and CMA-ES perform better than other methods. Therefore, the fuzzy algorithm reaches better $\bar{f}$ values in 4 from 5 different functions. This fact confirms that the fuzzy method is able to produce more accurate solutions than its competitors. From the analysis of σ_f in Table 8.13, it is clear that our fuzzy method obtain a better consistency than the other algorithms, since its produced solutions present a small dispersion. As it can be expected, the only exception is function f_{19}, where the fuzzy algorithm does not achieve the best performance Additional to such

Table 8.11 Minimization results of multimodal functions from Table 8.18 with $n = 100$

		HS	BAT	DE	PSO	CMA-ES	ABC	FUZZY
f_8	$\bar{f}$	−7.87E+03	−4.47E+03	−2.34E+04	−1.48E+04	−8.66E+03	−3.51E+04	**−1.62E+05**
	σ_f	584.340059	799.31231	3914.77561	1729.15916	831.547363	**554.101741**	47514.2941
	f_{Best}	−9141.81789	−6497.01004	−32939.624	−19362.7097	−10177.5382	−36655.8271	**−272938.668**
	f_{Worst}	−6669.02042	−2858.38997	−18358.9463	−12408.7377	−7375.40434	−33939.9476	**−87985.343**
f_9	$\bar{f}$	1.44E+03	9.12E+02	3.98E+02	7.49E+02	2.37E+02	6.49E+01	**4.00E−05**
	σ_f	41.464352	60.4785656	67.0536747	67.5932069	209.403173	7.90879809	**0.00015207**
	f_{Best}	1293.19918	772.978721	205.209989	635.167862	74.8466353	49.9922666	**0**
	f_{Worst}	1503.0147	1033.80416	522.504376	865.893944	806.576683	79.9136576	**0.0005994**
f_{10}	$\bar{f}$	2.06E+01	1.98E+01	2.42E+00	1.33E+01	6.14E−04	3.01E+00	**3.96E−12**
	σ_f	0.06052521	0.08670819	0.82882063	0.87767923	9.6205E−05	0.31154878	**2.1504E−11**
	f_{Best}	2.05E+01	1.96E+01	1.12E+00	1.16E+01	4.14E−04	2.26E+00	**1.51E−14**
	f_{Worst}	2.08E+01	2.00E+01	4.63E+00	1.51E+01	8.49E−04	3.59E+00	**1.18E−10**
f_{11}	$\bar{f}$	1.96E+03	2.38E+03	4.46E+00	1.18E+02	1.20E−03	1.61E−01	**0.00E+00**
	σ_f	81.7177655	134.986806	2.67096611	22.637086	0.00028091	0.14553457	**0**
	f_{Best}	1773.68789	2035.67623	1.02086153	86.1907763	0.00066065	0.01309143	**0**
	f_{Worst}	2083.84569	2557.11279	12.932624	170.763675	2.25E−03	0.64617359	**0**
f_{12}	$\bar{f}$	1.88E+09	2.55E+09	4.85E+04	2.06E+04	**2.20E−06**	5.96E−03	2.67E+02
	σ_f	110307213	242921094	140231.648	50859.5682	**7.1926E−07**	0.01533206	1423.14186
	f_{Best}	1.64E+09	1.87E+09	1.86E+00	2.93E+01	**1.27E−06**	2.91E−05	4.21E−01
	f_{Worst}	2090764763	2966731722	769582.376	251450.989	**4.07E−06**	0.06229125	7801.38816

(continued)

Table 8.11 (continued)

		HS	BAT	DE	PSO	CMA-ES	ABC	FUZZY
f_{13}	$\bar{f}$	3.54E+09	4.83E+09	6.63E+05	1.77E+06	2.20E+01	4.74E−03	**4.71E−05**
	σ_f	255789666	477261470	1076152.4	1373210.26	33.1622741	0.00582626	**1.2473E−05**
	f_{Best}	2936347707	3998731504	3147.16418	417561.443	9.98142144	0.00075547	**2.65E−05**
	f_{Worst}	3928279153	5901118241	4651437.83	7741055.37	176.314279	0.02480097	**7.86E−05**
f_{14}	$\bar{f}$	−1.57E+03	−1.82E+03	**−3.83E+03**	−2.46E+03	−3.57E+03	−3.78E+03	−2.30E+03
	σ_f	79.1693639	76.7852589	36.6930452	97.0301039	59.5947211	**22.8558428**	65.2812493
	f_{Best}	−1722.36681	−1945.12523	**−3900.49741**	−2683.50044	−3676.29234	−3823.95029	−2455.01558
	f_{Worst}	−1432.17812	−1695.67279	**−3740.63055**	−2280.67655	−3393.55796	−3724.2658	−2156.74064

Bold elements represent the best values

Table 8.12 *p*-values produced by Wilcoxon test comparing FUZZY versus HS, FUZZY versus BAT, FUZZY versus DE, FUZZY versus PSO, FUZZY versus CMA-ES and FUZZY versus ABC over the "averaged best fitness values" from Table 8.11

Wilcoxon test for Multimodal functions of Table 8.18 with n = 100						
FUZZY versus	HS	BAT	DE	PSO	CMA-ES	ABC
f_8	8.1340 −05▲	6.4720E −05▲	4.6920E −05▲	3.1664E −05▲	7.1163E −05▲	3.7920E −05▲
f_9	1.3642E −07▲	9.4982E −06▲	7.6012E −06▲	8.6620E −06▲	5.4901E −06▲	3.1362E −06▲
f_{10}	6.9482E −07▲	6.3482E −07▲	3.5698E −06▲	6.1345E −06▲	3.1692E −04▲	3.9302E −06▲
f_{11}	4.9842E −07▲	8.1647E −07▲	7.1352E −04▲	5.3120E −06▲	2.0162E −03▲	9.4867E −03▲
f_{12}	4.2682E −08▲	7.6801E −08▲	8.4672E −07▲	7.4682E −07▲	3.0521E −06▼	1.6428E −05▲
f_{13}	4.5926E −09▲	6.4720E −09▲	6.1680E −07▲	7.4682E −08▲	9.1722E −06▲	7.4682E −04▲
f_{14}	8.1550E −05▲	8.9647E −05▲	6.4923E −04▼	9.4212E −03▲	5.4682E −04▲	6.0125E −04▲
▲	7	7	6	7	6	7
▼	0	0	1	0	1	0
►	0	0	0	0	0	0

results, Table 8.13 shows that the Fuzzy method attains the best produced solution f_{Best} during the 30 independent executions than the other algorithms, except for f_{19} function. Besides, the worst fitness values f_{Worst} generated by the fuzzy technique maintain a better solution quality than the other methods excluding function f_{19}. The case of obtaining the best f_{Best} and f_{Worst} indexes reflexes the remarkable capacity of the fuzzy method to produce better solutions through use an efficient search strategy.

Table 8.14 shows the results of the Wilcoxon analysis over the averaged best fitness values $\bar{f}$ from Table 8.13. They indicate that the Fuzzy method performs better than the HS, BAT, DE and PSO algorithms. In the case of FUZZY versus CMA-ES, the FUZZY method maintains a better performance in all test functions except in problem f_{19}, where the CMA-ES produces better results than the FUZZY method. However, in the comparison between the FUZZY algorithm and ABC, FUZZY obtains the best results in all test functions except in problems f_{18} and f_{16}, where there is no statistical difference between the two methods.

In addition to the test in 50 dimensions, a second set of experiments have also conducted in 100 dimensions considering the same set of hybrid functions. Tables 8.15 and 8.16 present the results of the analysis in 100 dimensions. In Table 8.15, the indexes $\bar{f}$, σ_f, f_{Best} and f_{Worst}, obtained during the total number of executions, are reported. On the other hand, Table 8.14 presents the statistical Wilcoxon analysis of the averaged best fitness values $\bar{f}$ from Table 8.15.

Table 8.13 Minimization results of hybrid functions from Table 8.19 with n = 50

		HS	BAT	DE	PSO	CMA-ES	ABC	FUZZY
f_{15}	$\bar{f}$	7.9969E+13	5.1022E+21	12.3509776	9.64E+03	6.36E−06	5.23E−04	**2.49E−15**
	σ_f	1.3868E+14	2.7945E+22	16.9157032	5195.1729	2.0915E−06	0.00018266	**6.6512E−15**
	f_{Best}	1.5309E+10	8.1918E+12	0.01189652	4.19E+03	4.16E−06	2.68E−04	**3.17E−58**
	f_{Worst}	4.7627E+14	1.53E+23	65.8109321	2.50E+04	1.48E−05	1.02E−03	**2.53E−14**
f_{16}	$\bar{f}$	2706.73644	3508.41961	73.699317	5.99E+02	5.88E+02	4.91E+01	**4.90E+01**
	σ_f	103.253645	252.48393	13.915609	114.611979	9.03874746	0.19971751	**0.00025806**
	f_{Best}	2491.6759	2883.46006	49.0005972	393.824001	**48.9964485**	48.998348	48.9973691
	f_{Worst}	2876.20757	3869.21201	103.690098	878.998232	81.0717712	49.6854294	**48.9981462**
f_{17}	$\bar{f}$	1151151029	2105822689	67405.5759	3.13E+05	**5.40E+01**	8.96E+02	**5.40E+01**
	σ_f	113601255	190215425	193321.347	421404.743	0.00020018	132.212771	**9.8857E−05**
	f_{Best}	909668213	1637035871	413.088353	15529.3525	53.9999308	546.710586	**53.9998073**
	f_{Worst}	1428940501	2452560936	936409.163	1947952.34	54.0007602	1155.9351	**54.000177**
f_{18}	$\bar{f}$	2.0155E+14	3.3427E+19	54.2557544	9.06E+02	5.99E+01	**4.90E+01**	**4.90E+01**
	σ_f	3.9073E+14	1.83E+20	10.4970565	291.082635	12.4089453	0.01211373	**0**
	f_{Best}	911061731	3.2735E+13	49.0002421	530.607446	49.0000116	49.0021079	**49**
	f_{Worst}	1.85E+15	1.00E+21	97.123625	1597.0748	86.0099556	49.0607764	**49**
f_{19}	$\bar{f}$	7.7416E+14	7.7174E+18	−19.5833354	1.17E+06	**−1.44E+02**	−1.43E+02	2.18E+01
	σ_f	1.6757E+15	4.2201E+19	117.854019	5835860.88	0.39733093	**0.29998167**	472.608012
	f_{Best}	1343488881	1.939E+10	−143.748394	4511.91824	**−144.056723**	−143.608756	−83.2609165
	f_{Worst}	7.53E+15	2.31E+20	334.753741	32053489.3	**−142.208256**	−143.0037	2523.59236

Bold elements represent the best values

Table 8.14 *p*-values produced by Wilcoxon test comparing FUZZY versus HS, FUZZY versus BAT, FUZZY versus DE, FUZZY versus PSO, FUZZY versus CMA-ES and FUZZY versus ABC over the "averaged best fitness values" from Table 8.13

Wilcoxon test for Hybrid functions of Table 8.19 with $n = 50$						
FUZZY versus	HS	BAT	DE	PSO	CMA-ES	ABC
f_{15}	4.6102E−10▲	7.6801E−11▲	8.1253E−07▲	3.1678E−09▲	1.3542E−04▲	5.6932E−05▲
f_{16}	5.6922E−08▲	6.4982E−09▲	6.3142E−04▲	8.4320E−05▲	6.4931E−05▲	0.1560►
f_{17}	8.6523E−11▲	9.4685E−11▲	6.3352E−09▲	7.3477E−10▲	0.0956►	4.6501−04▲
f_{18}	3.4962E−11▲	7.6851E−12▲	4.6820E−04▲	5.3102E−06▲	4.8235E−04▲	0.1986►
f_{19}	7.6301E−10▲	9.3114E−11▲	4.3301E−07▲	6.0021E−09▲	6.3315E−07▼	5.8937E−07▼
▲	5	5	5	5	4	3
▼	0	0	0	0	1	1
►	0	0	0	0	0	1

Table 8.15 confirms the advantage of the Fuzzy method over HS, BAT, DE, PSO, CMA-ES and ABC. After analyzing the results, it is clear that the fuzzy method produces better results than HS, BAT, DE, PSO, CMA-ES and ABC in functions f_{15}–f_{18}. However, it can be seen that the Fuzzy method performs worse than CMA-ES and ABC in function f_{19}. Similar to the case of 50 dimensions, in the experiments of 100 dimensions, the fuzzy algorithm obtains solutions with the smallest level of dispersion (σ_f). This consistency is valid for all functions, except for problem f_{19}, where the CMA-ES obtain the best σ_f value. Considering the f_{Best} and f_{Worst} indexes, similar conclusion can be established that in the case of 50 dimensions. In 100 dimension, it is also observed that the fuzzy technique surpass all algorithms in the production of high quality solutions.

On the other hand, the data obtained from the Wilcoxon analysis (Table 8.16) demonstrates that the FUZZY method performs better than the other metaheuristic algorithms in all test functions, except in problem f_{18}, where the CMA-ES and ABC produce the best results. In the Table 8.16, it is also summarized the results of the analysis through the symbols ▲, ▼, and ►. The conclusions of the Wilcoxon test statistically validate the results of Table 8.15. They indicate that the superior performance of the fuzzy method is as a consequence of a better search strategy and not for random effects.

Table 8.15 Minimization results of hybrid functions from Table 8.19 with n = 100

		HS	BAT	DE	PSO	CMA-ES	ABC	FUZZY
f_{15}	$\bar{f}$	1.07E+38	1.45E+46	1.67E+02	3.58E+04	8.48E−03	8.13E−02	**1.03E−05**
	σ_f	2.6648E+38	7.9319E+46	212.77092	19777.8798	0.00248591	0.04144673	**5.64261E−05**
	f_{Best}	1.88E+28	1.31E+37	5.17E+00	1.58E+04	5.99E−03	4.35E−02	**4.50E−44**
	f_{Worst}	1.21E+39	4.34E+47	1.15E+03	9.15E+04	1.75E−02	2.66E−01	**3.09E−04**
f_{16}	$\bar{f}$	6.46E+03	8.00E+03	1.90E+02	1.36E+03	1.55E+02	1.81E+02	**9.90E+01**
	σ_f	316.896114	426.761823	28.0273818	129.05773	18.161094	18.5185506	**0.00187142**
	f_{Best}	5753.69747	7125.84623	148.521771	1116.8234	122.010095	129.807203	**98.9958572**
	f_{Worst}	6997.63377	8965.08163	260.542179	1681.54333	187.628218	206.634292	**99.0057217**
f_{17}	$\bar{f}$	3.57E+09	5.03E+09	6.85E+05	2.10E+06	5.95E+02	3.72E+03	**1.09E+02**
	σ_f	251070990	401513619	1158365.05	1426951.79	84.33472	459.761456	**0.0026802**
	f_{Best}	2908492728	4028081811	4767.47553	428429.323	428.761773	2973.30401	**108.99997**
	f_{Worst}	3953742523	5605713616	4789383.94	6274361.22	784.972289	4756.54052	**109.01469**
f_{18}	$\bar{f}$	3.30E+38	5.29E+43	1.33E+02	2.09E+03	1.49E+02	3.34E+02	**1.08E+02**
	σ_f	1.2038E+39	2.669E+44	20.7615151	481.839456	23.6851947	908.915277	**8.76059629**
	f_{Best}	5.02E+29	4.38E+34	1.01E+02	1.24E+03	1.12E+02	**9.90E+01**	99.6507148
	f_{Worst}	5.86E+39	1.47E+45	1.92E+02	3.51E+03	2.10E+02	4.24E+03	**134.545048**
f_{19}	$\bar{f}$	1.01E+38	1.49E+44	9.38E+03	6.45E+07	**−2.94E+02**	−2.01E+02	4.29E+07
	σ_f	3.9907E+38	6.2388E+44	40545.7303	217095412	**0.55047399**	1.27097202	200132558
	f_{Best}	4.34E+29	1.43E+36	−1.10E+02	4.71E+04	−2.95E+02	**−295.68221**	−9.45E+01
	f_{Worst}	2.18E+39	3.36E+45	2.23E+05	1.17E+09	**−2.92E+02**	−288.145624	1.08E+09

Bold elements represent the best values

Table 8.16 *p*-values produced by Wilcoxon test comparing FUZZY versus HS, FUZZY versus BAT, FUZZY versus DE, FUZZY versus PSO, FUZZY versus CMA-ES and FUZZY versus ABC over the "averaged best fitness values" from Table 8.15

Wilcoxon test for Hybrid functions of Table 8.19 with $n = 100$						
FUZZY versus	HS	BAT	DE	PSO	CMA-ES	ABC
f_{15}	8.4682E −12▲	9.7624E −12▲	6.4950E −07▲	7.0012E −08▲	3.1261E −04▲	7.6823E −04▲
f_{16}	7.6332E −05▲	8.4220E −05▲	6.5010E −04▲	2.0035E −04▲	3.9630E −04▲	6.0012E −11▲
f_{17}	5.3422E −08▲	6.8892E −08▲	3.3019E −07▲	9.0394E −07▲	4.2961E −04▲	8.6301E −05▲
f_{18}	4.9302E −12▲	8.3670E −12▲	5.6312E −04▲	3.4621E −05▲	6.0341E −04▲	1.3025E −05▲
f_{19}	6.9210E −11▲	2.4950E −12▲	6.3301E −06▲	2.0182E −04▲	6.3019E −07▼	4.1305E −07▼
▲	5	5	5	5	4	4
▼	0	0	0	0	1	1
►	0	0	0	0	0	0

8.5.2.4 Convergence Experiments

The comparison of the final fitness value cannot completely describe the searching performance of an optimization algorithm. Therefore, in this section, a convergence test on the seven compared algorithms has been conducted. The purpose of this experiment is to evaluate the velocity with which a compared method reaches the optimum. In the experiment, the performance of each algorithm is considered over all functions (f_1–f_{19}) from Appendix, operated in 50 dimensions. In order to build the convergence graphs, we employ the raw simulation data generated in Sects. 8.5.2.1, 8.5.2.2 and 8.5.2.3. As each function is executed 30 times for each algorithm, we select the convergence data of the run which represents the median final result. Figures 8.5, 8.6 and 8.7 show the convergence data of the seven compared algorithms. Figure 8.5 presents the convergence results for functions f_1–f_6, Fig. 8.6 for functions f_7–f_{12} and Fig. 8.7 for functions f_{13}–f_{19}. In the figures, the x-axis is the elapsed function evaluations, and the y-axis represents the best fitness values found.

From Fig. 8.5, it is clear that the fuzzy method presents a better convergence than the other algorithms for functions f_1, f_2, f_4 and f_5. However, for function f_3 and f_6 the CMA-ES reaches faster an optimal value. After an analysis of Fig. 8.5, we can say that the Fuzzy method and the CMA-ES algorithm attain the best convergence responses whereas the other techniques maintain slower responses. In Fig. 8.6, the convergence graphs show that the fuzzy method obtains the best responses for functions f_9, f_{10} and f_{11}. In function f_7, even though the Fuzzy technique finds in a fat way optimal solutions, the DE algorithm presents the best convergence result. An interesting case is function f_9, where several optimization methods such as FUZZY, CMA-ES, ABC and DE obtain an acceptable convergence response. In case of function f_8, the DE and ABC methods own the best convergence properties. Finally, in function f_{12}, the CMA-ES attains the fastest reaction. Finally, in Fig. 8.7, the convergence responses for functions f_{13}–f_{19} are presented. In function f_{13} of Fig. 8.7, the algorithms CMA-ES and ABC obtain the best responses. In case of function f_{14}, DE and ABC find an optimal solution in a prompt way than the other optimization techniques. Although for functions f_{15}–f_{18} the fuzzy algorithm reaches the fastest convergence reaction, the CMA-ES method maintains a similar response. For function f_{19}, the CMA-ES and ABC own the best convergence properties.

Therefore, the convergence speed of the fuzzy method in solving unimodal optimization (f_1–f_7) problems is faster than HS, BAT, DE, PSO, CMA-ES and ABC, except in f_7, where the CMA-ES reaches the best response. On the other hand, when solving multimodal optimization problems (f_8–f_{14}), the fuzzy algorithm generally converges as fast as or even faster than the compared algorithms. This phenomenon can be clearly observed in Figs. 8.6 and 8.7, where the method generates a similar convergence curve to the others, even in the worst case scenario. Finally, after analyzing the performance of all algorithms on hybrid functions (f_{15}–f_{19}), it is clear that the convergence response of the approach is not as fast as CMA-ES. In fact, the fuzzy and the CMA-ES algorithms present the best convergence properties when they face the optimization of hybrid functions.

8.5.2.5 Computational Complexity

In this section, the computational complexity of all methods is evaluated. Metaheuristic methods are, in general, complex processes with several random operations and stochastic sub-routines. Therefore, it is impractical to conduct a complexity analysis from a deterministic point of view. For that reason, the computational complexity (C) is used in order to evaluate the computational effort of each algorithm. C exhibits the averaged CPU time invested by an algorithm with regard to a common time reference, when it is under operation. In order to assess the computational complexity, the procedure presented in [48] has been conducted. Under this process, C is obtained through the subsequent method:

1. The time reference T_0 is computed. T_0 represents the computing time consumed by the execution of the following standard code:

```
for j=1:1000000
v=0.55+j
v=v+v; v=v/2; v=v*v; v=sqrt(v); v=exp(v); v=v/(v+2);
end
```

(continued)

(continued)

2.	Evaluate the computing time T_1 for function operation. T_1 Expresses the time in which 200,000 runs of function f_9 (multimodal) are executed (only the function without optimization method). In the test, the function f_9 is operated with $n = 100$
3.	Calculate the execution time T_2 for the optimization algorithm. T_2 exhibits the elapsed time in which 200,000 function evaluations of f_9 are executed (here, optimization method and function are combined)
4.	The average time $\bar{T}_2$ is computed. First, execute the Step 3 five times. Then, extract their average value $\bar{T}_2 = (T_2^1 + T_2^2 + T_2^3 + T_2^4 + T_2^5)/5$
5.	The computational complexity C is obtained as follows: $C = (\bar{T}_2 - T_1)/T_0$

Under this process, the computational complexity (C) values of HS, BAT, DE, PSO, CMA-ES, ABC, and FUZZY are obtained. Their values correspond to 77.23, 81.51, 51.20, 36.87, 40.77, 70.17 and 40.91, respectively. A smaller C value indicates that the method is less complex, which allows a faster execution speed under the same evaluation conditions. An analysis of the experiment results shows that although the FUZZY algorithm is slightly more complex than PSO and CMA-ES, their computational complexity (C) values are comparable. Additionally, the three algorithms are significantly less computationally complex than HS, BAT, DE and ABC.

8.6 Conclusions

Recently, several new metaheuristic algorithms have been proposed with interesting results. Most of them use operators based on metaphors of natural or social elements to evolve candidate solutions. Although humans have demonstrated their potential to solve real-life complex optimization problems, the use of human knowledge to build optimization algorithms has been less popular than the natural or social metaphors. In this chapter, a methodology to implement human-knowledge-based optimization strategies has been presented. Under the approach, a conducted search strategy is modeled in the rule base of a Takagi-Sugeno Fuzzy inference system, so that the implemented fuzzy rules express the conditions under which candidate solutions are evolved during the optimization process.

All the approaches reported in the literature that integrate Fuzzy logic and metaheuristic techniques consider the optimization capabilities of the metaheuristic algorithms for improving the performance of fuzzy systems. In this method, the approach is completely different. Under this new schema, the Fuzzy system directly conducts the search strategy during the optimization process. In this chapter our intent is to propose a methodology for emulating human search strategies in an algorithmic structure. To the best of our knowledge, this is the first time that a fuzzy system is used as a metaheuristic algorithm.

The Fuzzy methodology presents three important characteristics: (1) *Generation.* Under this methodology, fuzzy logic provides a simple and well-known method for

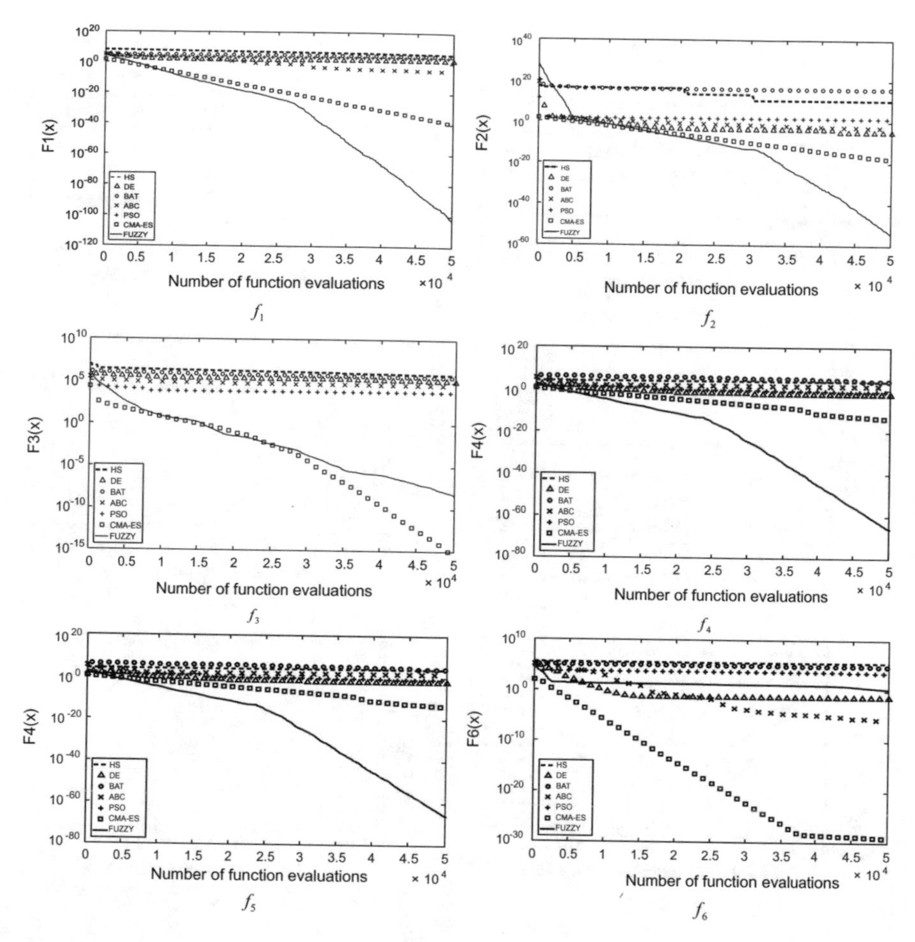

Fig. 8.5 Convergence test results for functions $f_1 - f_6$

constructing a search strategy via the use of human knowledge. (2) *Transparency*. It generates fully interpretable models whose content expresses the search strategy as humans can conduct it. (3) *Improvement*. As human experts interact with an optimization process, they obtain a better understanding of successful search strategies capable of finding optimal solutions. As a result, new rules are added so that their inclusion in the existing rule base improves the quality of the original search strategy. Under the Fuzzy methodology, new rules can be easily incorporated to an already existent system. The addition of such rules allows the capacities of the original system to be extended.

To demonstrate the ability and robustness of our approach, the fuzzy algorithm has been experimentally evaluated with a test suite of 19 benchmark functions. To assess the performance of the fuzzy algorithm, it has been compared to other popular optimization approaches based on evolutionary principles currently in use.

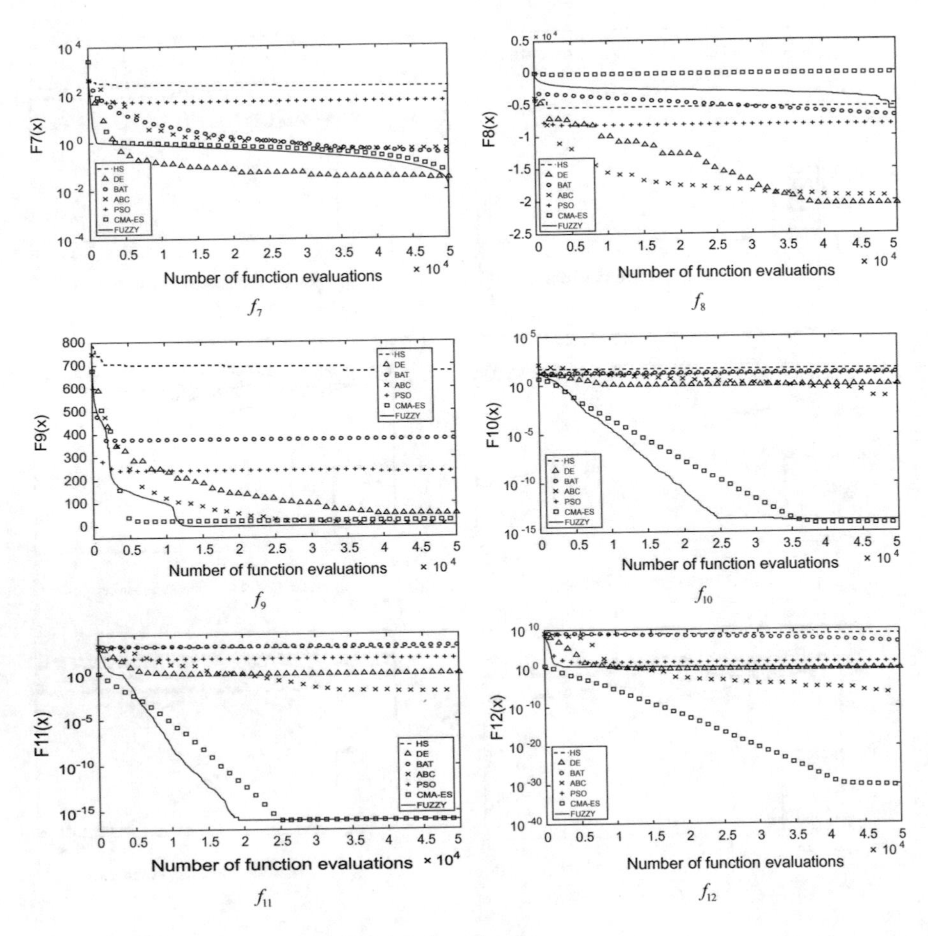

Fig. 8.6 Convergence test results for functions f_7– f_{12}

The results, statistically validated, have confirmed that the Fuzzy algorithm outperforms its competitors for most of the test functions in terms of its solution quality and convergence.

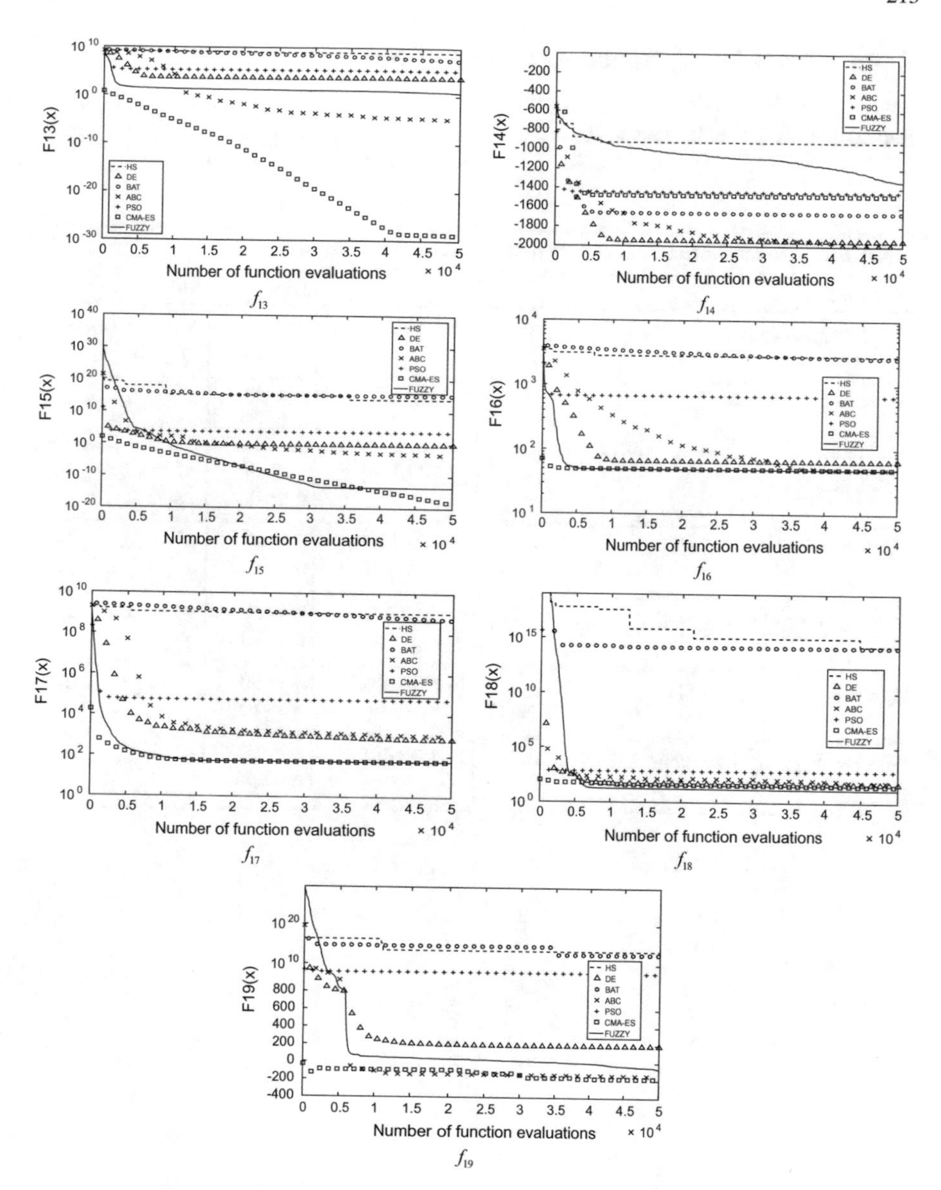

Fig. 8.7 Convergence test results for functions f_{13}–f_{19}

Appendix: List of Benchmark Functions

See Tables 8.17, 8.18 and 8.19.

Table 8.17 Unimodal test functions used in the experimental study

Function	S	Dim	Minimum
$f_1(\mathbf{x}) = \sum_{i=1}^{n} x_i^2$	$[-100,100]^n$	$n = 50$ $n = 100$	$\mathbf{x}^* = (0,\ldots,0)$; $f(\mathbf{x}^*) = 0$
$f_2(\mathbf{x}) = \sum_{i=1}^{n} \lvert x_i \rvert + \prod_{i=1}^{n} \lvert x_i \rvert$	$[-10,10]^n$	$n = 50$ $n = 100$	$\mathbf{x}^* = (0,\ldots,0)$; $f(\mathbf{x}^*) = 0$
$f_3(\mathbf{x}) = \sum_{i=1}^{n} \left(\sum_{j=1}^{i} x_j\right)^2$	$[-100,100]^n$	$n = 50$ $n = 100$	$\mathbf{x}^* = (0,\ldots,0)$; $f(\mathbf{x}^*) = 0$
$f_4(\mathbf{x}) = \max_i\{\lvert x_i \rvert, 1 \le i \le n\}$	$[-100,100]^n$	$n = 50$ $n = 100$	$\mathbf{x}^* = (0,\ldots,0)$; $f(\mathbf{x}^*) = 0$
$f_5(\mathbf{x}) = \sum_{i=1}^{n-1} \left[100(x_{i+1} - x_i^2)^2 + (x_i - 1)^2\right]$	$[-30,30]^n$	$n = 50$ $n = 100$	$\mathbf{x}^* = (1,\ldots,1)$; $f(\mathbf{x}^*) = 0$
$f_6(\mathbf{x}) = \sum_{i=1}^{n} (\lfloor x_i + 0.5 \rfloor)^2$	$[-100,100]^n$	$n = 50$ $n = 100$	$\mathbf{x}^* = (0,\ldots,0)$; $f(\mathbf{x}^*) = 0$
$f_7(\mathbf{x}) = \sum_{i=1}^{n} i x_i^4 + random(0,1)$	$[-1.28,1.28]^n$	$n = 50$ $n = 100$	$\mathbf{x}^* = (0,\ldots,0)$; $f(\mathbf{x}^*) = 0$

Table 8.18 Multimodal test functions used in the experimental study

Function	Range	Dimension	Optimum
$f_8(\mathbf{x}) = \sum_{i=1}^{n} -x_i \sin\left(\sqrt{\lvert x_i \rvert}\right)$	$[-500,500]^n$	$n = 50$ $n = 100$	$\mathbf{x}^* = (420, \ldots, 420)$; $f(\mathbf{x}^*) = -418.9829 \times n$
$f_9(\mathbf{x}) = \sum_{i=1}^{n} \left[x_i^2 - 10\cos(2\pi x_i) + 10\right]$	$[-5.12,5.12]^n$	$n = 50$ $n = 100$	$\mathbf{x}^* = (0, \ldots, 0)$; $f(\mathbf{x}^*) = 0$
$f_{10}(\mathbf{x}) = -20\exp\left(-0.2\sqrt{\frac{1}{n}\sum_{i=1}^{n} x_i^2}\right)$ $- \exp\left(\frac{1}{n}\sum_{i=1}^{n} \cos(2\pi x_i)\right) + 20 + \exp$	$[-32,32]^n$	$n = 50$ $n = 100$	$\mathbf{x}^* = (0, \ldots, 0)$; $f(\mathbf{x}^*) = 0$
$f_{11}(\mathbf{x}) = \frac{1}{4000}\sum_{i=1}^{n} x_i^2 - \prod_{i=1}^{n} \cos\left(\frac{x_i}{\sqrt{i}}\right) + 1$	$[-600,600]^n$	$n = 50$ $n = 100$	$\mathbf{x}^* = (0, \ldots, 0)$; $f(\mathbf{x}^*) = 0$
$f_{12}(\mathbf{x}) = \frac{\pi}{n}\left\{10\sin(\pi y_1) + \sum_{i=1}^{n-1} (y_i - 1)^2\left[1 + 10\sin^2(\pi y_{i+1})\right]\right.$ $\left. + (y_n - 1)^2\right\} + \sum_{i=1}^{n} u(x_i, 10, 100, 4)$ $y_i = 1 + \frac{(x_i+1)}{4}$ $u(x_i, a, k, m) = \begin{cases} k(x_i - a)^m & x_i > a \\ 0 & -a \le x_i \le a \\ k(-x_i - a)^m & x_i < a \end{cases}$	$[-50,50]^n$	$n = 50$ $n = 100$	$\mathbf{x}^* = (0, \ldots, 0)$; $f(\mathbf{x}^*) = 0$
$f_{13}(\mathbf{x}) = 0.1\left\{\sin^2(3\pi x_1) + \sum_{i=1}^{n} (x_i - 1)^2\right.$ $\left.\left[1 + \sin^2(3\pi x_i + 1)\right] + (x_n - 1)^2\left[1 + \sin^2(2\pi x_n)\right]\right\}$ $+ \sum_{i=1}^{n} u(x_i, 5, 100, 4)$; $u(x_i, a, k, m) = \begin{cases} k(x_i - a)^m & x_i > a \\ 0 & -a < x_i < a \\ k(-x_i - a)^m & x_i < -a \end{cases}$	$[-10,10]^n$	$n = 50$ $n = 100$	$\mathbf{x}^* = (1, \ldots, 1)$; $f(\mathbf{x}^*) = 0$
$f_{14}(x) = \frac{1}{2}\sum_{i=1}^{n} \left(x_i^4 - 16x_i^2 + 5x_i\right)$	$[-5,5]^n$	$n = 50$ $n = 100$	$\mathbf{x}^* = (-2.90, \ldots, -2.90)$; $f(\mathbf{x}^*) = -39.16599 \times n$

Table 8.19 Hybrid test functions used in the experimental study

Function	S	Dim	Minimum
$f_{15}(\mathbf{x}) = f_1(\mathbf{x}) + f_2(\mathbf{x}) + f_9(\mathbf{x})$	$[-100,100]^n$	$n = 50$ $n = 100$	$\mathbf{x}^* = (0,\ldots,0)$; $f(\mathbf{x}^*) = 0$
$f_{16}(\mathbf{x}) = f_9(\mathbf{x}) + f_5(\mathbf{x}) + f_{11}(\mathbf{x})$	$[-100,100]^n$	$n = 50$ $n = 100$	$\mathbf{x}^* = (0,\ldots,0)$; $f(\mathbf{x}^*) = n - 1$
$f_{17}(\mathbf{x}) = f_3(\mathbf{x}) + f_5(\mathbf{x}) + f_{10}(\mathbf{x}) + f_{13}(\mathbf{x})$	$[-100,100]^n$	$n = 50$ $n = 100$	$\mathbf{x}^* = (0,\ldots,0)$; $f(\mathbf{x}^*) = (1.1 \times n) - 1$
$f_{18}(\mathbf{x}) = f_2(\mathbf{x}) + f_5(\mathbf{x}) + f_9(\mathbf{x}) + f_{10}(\mathbf{x}) + f_{11}(\mathbf{x})$	$[-100,100]^n$	$n = 50$ $n = 100$	$\mathbf{x}^* = (0,\ldots,0)$; $f(\mathbf{x}^*) = n - 1$
$f_{19}(\mathbf{x}) = f_1(\mathbf{x}) + f_2(\mathbf{x}) + f_8(\mathbf{x}) + f_{10}(\mathbf{x}) + f_{12}(\mathbf{x})$	$[-100,100]^n$	$n = 50$ $n = 100$	$\mathbf{x}^* = (1,\ldots,1)$; $f(\mathbf{x}^*) = 0$

References

1. Zadeh, L.A.: Fuzzy sets. Inf. Control **8**, 338–353 (1965)
2. He, Y., Chen, H., He, Z., Zhou, L.: Multi-attribute decision making based on neutral averaging operators for intuitionistic fuzzy information. Appl. Soft Comput. **27**, 64–76 (2015)
3. Taur, J., Tao, C.W.: Design and analysis of region-wise linear fuzzy controllers. IEEE Trans. Syst. Man Cybern. B Cybern. **27**(3), 526–532 (1997)
4. Ali, M.I., Shabir, M.: Logic connectives for soft sets and fuzzy soft sets. IEEE Trans. Fuzzy Syst. **22**(6), 1431–1442 (2014)
5. Novák, V., Hurtík, P., Habiballa, H., Štepnička, M.: Recognition of damaged letters based on mathematical fuzzy logic analysis. J. Appl. Logic **13**(2), 94–104 (2015)
6. Papakostas, G.A., Hatzimichailidis, A.G., Kaburlasos, V.G.: Distance and similarity measures between intuitionistic fuzzy sets: a comparative analysis from a pattern recognition point of view. Pattern Recogn. Lett. **34**(14), 1609–1622 (2013)
7. Wang, X., Fu, M., Ma, H., Yang, Y.: Lateral control of autonomous vehicles based on fuzzy logic. Control Eng. Pract. **34**, 1–17 (2015)
8. Castillo, O., Melin, P.: A review on interval type-2 fuzzy logic applications in intelligent control. Inf. Sci. **279**, 615–631 (2014)
9. Raju, G., Nair, M.S.: A fast and efficient color image enhancement method based on fuzzy-logic and histogram. AEU Int. J. Electron. Commun. **68**(3), 237–243 (2014)
10. Zareiforoush, H., Minaei, S., Alizadeh, M.R., Banakar, A.: A hybrid intelligent approach based on computer vision and fuzzy logic for quality measurement of milled rice. Measurement **66**, 26–34 (2015)
11. Nanda, S.J., Panda, G.: A survey on nature inspired metaheuristic algorithms for partitional clustering. Swarm Evol. Comput. **16**, 1–18 (2014)
12. Kennedy, J., Eberhart, R.: Particle swarm optimization. In: Proceedings of the 1995 IEEE International Conference on Neural Networks, vol. 4, pp. 1942–1948, December 1995
13. Karaboga, D.: An idea based on honey bee swarm for numerical optimization. Technical Report-TR06. Engineering Faculty, Computer Engineering Department, Erciyes University (2005)
14. Geem, Z.W., Kim, J.H., Loganathan, G.V.: A new heuristic optimization algorithm: harmony search. Simulations **76**, 60–68 (2001)

15. Yang, X.S.: A new metaheuristic bat-inspired algorithm. In: Cruz, C., González, J., Krasnogor, G.T.N., Pelta, D.A. (eds.) Nature Inspired Cooperative Strategies for Optimization (NISCO 2010), Studies in Computational Intelligence, vol. 284, pp. 65–74. Springer, Berlin (2010)
16. Yang, X.S.: Firefly algorithms for multimodal optimization. In: Stochastic Algorithms: Foundations and Applications, SAGA 2009, Lecture Notes in Computer Sciences, vol. 5792, pp. 169–178 (2009)
17. Cuevas, E., Cienfuegos, M., Zaldívar, D., Pérez-Cisneros, M.: A swarm optimization algorithm inspired in the behavior of the social-spider. Expert Syst. Appl. **40**(16), 6374–6384 (2013)
18. Cuevas, E., González, M., Zaldivar, D., Pérez-Cisneros, M., García, G.: An algorithm for global optimization inspired by collective animal behavior. Discrete Dyn. Nat. Soc. art. no. 638275 (2012)
19. Storn, R., Price, K.: Differential evolution—a simple and efficient adaptive scheme for global optimisation over continuous spaces. Technical Report TR-95–012. ICSI, Berkeley, CA (1995)
20. Goldberg, D.E.: Genetic Algorithm in Search Optimization and Machine Learning. Addison-Wesley, USA (1989)
21. Herrera, F.: Genetic fuzzy systems: taxonomy, current research trends and prospects. Evol. Intell. **1**, 27–46 (2008)
22. Fernández, A., López, V., del Jesus, M.J., Herrera, F.: Revisiting evolutionary fuzzy systems: taxonomy, applications, new trends and challenges. Knowl. Based Syst. **80**, 109–121 (2015)
23. Caraveo, C., Valdez, F., Castillo, O.: Optimization of fuzzy controller design using a new bee colony algorithm with fuzzy dynamic parameter adaptation. Appl. Soft Comput. **43**, 131–142 (2016)
24. Castillo, O., Neyoy, H., Soria, J., Melin, P., Valdez, F.: A new approach for dynamic fuzzy logic parameter tuning in ant colony optimization and its application in fuzzy control of a mobile robot. Appl. Soft Comput. **28**, 150–159 (2015)
25. Olivas, F., Valdez, F., Castillo, O., Melin, P.: Dynamic parameter adaptation in particle swarm optimization using interval type-2 fuzzy logic. Soft. Comput. **20**(3), 1057–1070 (2016)
26. Castillo, O., Ochoa, P., Soria, J.: Differential evolution with fuzzy logic for dynamic adaptation of parameters in mathematical function optimization. In: Imprecision and Uncertainty in Information Representation and Processing, pp. 361–374 (2016)
27. Guerrero, M., Castillo, O., García Valdez, M.: Fuzzy dynamic parameters adaptation in the cuckoo search algorithm using fuzzy logic. In: CEC 2015, pp. 441–448
28. Alcala, R., Gacto, M.J., Herrera, F.: A fast and scalable multiobjective genetic fuzzy system for linguistic fuzzy modeling in high-dimensional regression problems. IEEE Trans. Fuzzy Syst. **19**(4), 666–681 (2011)
29. Alcala-Fdez, J., Alcala, R., Gacto, M.J., Herrera, F.: Learning the membership function contexts for mining fuzzy association rules by using genetic algorithms. Fuzzy Sets Syst. **160** (7), 905–921 (2009)
30. Alcala, R., Alcala-Fdez, J., Herrera, F.: A proposal for the genetic lateral tuning of linguistic fuzzy systems and its interaction with rule selection. IEEE Trans. Fuzzy Syst. **15**(4), 616–635 (2007)
31. Alcala-Fdez, J., Alcala, R., Herrera, F.: A fuzzy association rule-based classification model for high-dimensional problems with genetic rule selection and lateral tuning. IEEE Trans. Fuzzy Syst. **19**(5), 857–872 (2011)
32. Carmona, C.J., Gonzalez, P., del Jesus, M.J., Navio-Acosta, M., Jimenez-Trevino, L.: Evolutionary fuzzy rule extraction for subgroup discovery in a psychiatric emergency department. Soft. Comput. **15**(12), 2435–2448 (2011)
33. Cordon, O.: A historical review of evolutionary learning methods for Mamdani-type fuzzy rule-based systems: designing interpretable genetic fuzzy systems. Int. J. Approximate Reasoning **52**(6), 894–913 (2011)
34. Cruz-Ramirez, M., Hervas-Martinez, C., Sanchez-Monedero, J., Gutierrez, P.A.: Metrics to guide a multi-objective evolutionary algorithm for ordinal classification. Neurocomputing **135**, 21–31 (2014)
35. Lessmann, S., Caserta, M., Arango, I.M.: Tuning metaheuristics: a data mining based approach for particle swarm optimization. Expert Syst. Appl. **38**(10), 12826–12838 (2011)

36. Sörensen, K.: Metaheuristics—the metaphor exposed. Int. Trans. Oper. Res. **22**(1), 3–18 (2015)
37. Omid, M., Lashgari, M., Mobli, H., Alimardani, R., Mohtasebi, S., Hesamifard, R.: Design of fuzzy logic control system incorporating human expert knowledge for combine harvester. Expert Syst. Appl. **37**(10), 7080–7085 (2010)
38. Fullér, R., Canós Darós, L., Darós, M.J.C.: Transparent fuzzy logic based methods for some human resource problems. Revista Electrónica de Comunicaciones y Trabajos de ASEPUMA **13**, 27–41 (2012)
39. Cordón, O., Herrera, F.: A three-stage evolutionary process for learning descriptive and approximate fuzzy-logic-controller knowledge bases from examples. Int. J. Approximate Reasoning **17**(4), 369–407 (1997)
40. Takagi, T., Sugeno, M.: Fuzzy identification of systems and its applications to modeling and control. IEEE Trans. Syst. Man Cybern. **SMC-15**, 116–132 (1985)
41. Mamdani, E., Assilian, S.: An experiment in linguistic synthesis with a fuzzy logic controller. Int. J. Man Mach. Stud. **7**, 1–13 (1975)
42. Bagis, A., Konar, M.: Comparison of Sugeno and Mamdani fuzzy models optimized by artificial bee colony algorithm for nonlinear system modelling. Trans. Inst. Measur. Control **38**(5), 579–592 (2016)
43. Guney, K., Sarikaya, N.: Comparison of mamdani and sugeno fuzzy inference system models for resonant frequency calculation of rectangular microstrip antennas. Prog. Electromagnet. Res. B **12**, 81–104 (2009)
44. Baldick, R.: Applied Optimization. Cambridge University Press, Cambridge (2006)
45. Simon, D.: Evolutionary Algorithms—Biologically Inspired and Population Based Approaches to Computer Intelligence. Wiley, USA (2013)
46. Wong, S.Y., Yap, K.S., Yap, H.J., Tan, S.C., Chang, S.W.: On equivalence of FIS and ELM for interpretable rule-based knowledge representation. IEEE Trans. Neural Networks Learn. Syst. **27**(7), 1417–1430 (2015)
47. Yap, K.S., Wong, S.Y., Tiong, S.K.: Compressing and improving fuzzy rules using genetic algorithm and its application to fault detection. In: IEEE 18th Conference on Emerging Technologies & Factory Automation (ETFA), vol. 1, pp. 1–4 (2013)
48. Liang, J.J., Qu, B.-Y., Suganthan, P.N.: Problem definitions and evaluation criteria for the CEC 2015, Special session and competition on single objective real parameter numerical optimization. Technical Report 201311. Computational Intelligence Laboratory, Zhengzhou University, Zhengzhou China and Nanyang Technological University, Singapore (2015)
49. Hansen, N., Ostermeier, A., Gawelczyk, A.: On the adaptation of arbitrary normal mutation distributions in evolution strategies: the generating set adaptation. In: Proceedings of the 6th International Conference on Genetic Algorithms, pp. 57–64 (1995)
50. Boussaïda, I., Lepagnot, J., Siarry, P.: A survey on optimization metaheuristics. Inf. Sci. **237**, 82–117 (2013)
51. Yu, J.J.Q., Li, V.O.K.: A social spider algorithm for global optimization. Appl. Soft Comput. **30**, 614–627 (2015)
52. Li, M.D., Zhao, H., Weng, X.W., Han, T.: A novel nature-inspired algorithm for optimization: virus colony search. Adv. Eng. Softw. **92**, 65–88 (2016)
53. Han, M., Liu, C., Xing, J.: An evolutionary membrane algorithm for global numerical optimization problems. Inf. Sci. **276**, 219–241 (2014)
54. Meng, Z., Pan, J.-S.: Monkey king evolution: a new memetic evolutionary algorithm and its application in vehicle fuel consumption optimization. Knowl. Based Syst. **97**, 144–157 (2016)
55. https://www.lri.fr/~hansen/cmaesintro.html
56. Wilcoxon, F.: Individual comparisons by ranking methods. Biometrics **1**, 80–83 (1945)
57. Garcia, S., Molina, D., Lozano, M., Herrera, F.: A study on the use of non-parametric tests for analyzing the evolutionary algorithms' behavior: a case study on the CEC '2005, Special session on real parameter optimization. J Heurist (2008). https://doi.org/10.1007/s10732-008-9080-4

Printed in the United States
By Bookmasters